recent advances in phytochemistry

volume 21

Phytochemical Effects of Environmental Compounds

RECENT ADVANCES IN PHYTOCHEMISTRY

Proceedings of the Phytochemical Society of North America
General Editor: Eric E. Conn, *University of California, Davis, California*

Recent Volumes in the Series

A Continuation Order Plan is available for this series. A continuation order will bring delivery of
each new volume immediately upon publication. Volumes are billed only upon actual shipment.
For further information please contact the publisher.

recent advances in phytochemistry

volume 21

Phytochemical Effects of Environmental Compounds

Edited by

**James A. Saunders
and Lynn Kosak-Channing**
Beltsville Agricultural Research Center
Beltsville, Maryland

and

Eric E. Conn
University of California, Davis
Davis, California

PLENUM PRESS • NEW YORK AND LONDON

ꭴꬴ

Library of Congress Cataloging in Publication Data

Phytochemical Society of North America. Meeting (26th: 1986: College Park, Md.)
   Phytochemical effects of environmental compounds.

   (Recent advances in phytochemistry; v. 21)
   "Proceedings of the Twenty-sixth Annual Meeting of the Phytochemical Society of
North America, held July 13–17, 1986, in College Park, Maryland"—Verso of t.p.
   Includes bibliographical references and index.
   1. Plants, Effect of pollution on—Congresses. 2. Plants, Effect of chemicals on—
Congresses. 3. Botanical chemistry—Congresses. 4. Environmental
chemistry—Congresses. I. Saunders, James A. II. Kosak-Channing, Lynn. III. Conn,
Eric E. IV. Title. V. Series.
QK861.R38 vol. 21   581.19′2 s   87-18659
[QK750]   [581.2′4]
ISBN 0-306-42675-7

Proceedings of the Twenty-sixth Annual Meeting of the Phytochemical Society
of North America, held July 13–17, 1986, in College Park, Maryland

© 1987 Plenum Press, New York
A Division of Plenum Publishing Corporation
233 Spring Street, New York, N.Y. 10013

Printed in the United States of America

12/18/87

PREFACE

The influence of compounds in the environment on the chemistry of plants is a topic which has economic and scientific implications of global importance. Selected presentations in this symposium covered several topics within this immense field, inclusive of air, soil, and aquatic sources of the compounds. As demonstrated in Chapter 4 by O'Keeffe et al. we have not restricted the discussion solely to negative aspects of anthropogenic compounds. Nor could we begin to cover comprehensively all major classes of environmental compounds in the air, soil or water that may have an effect on the phytochemistry of plants. Our intent was to focus on some of the timely and well publicized environmental constituents such as ozone, sulfur dioxide, acid rain, and others, to provide an authoritative publication specifically related to environmental modifications of plant chemistry.

The concept of this symposium originated with the Executive Committee of the Phytochemical Society of North America in 1983. It was brought to fruition during July 13-17, 1986 on the campus of the University of Maryland at the annual meeting of the PSNA through the efforts of the Symposium Committee composed of James A. Saunders and Lynn Kosak-Channing. Financial support for this meeting was provided by the Phytochemical Society of North America, as well as by generous contributions from E.I. du Pont de Nemours & Company and the U.S. Department of Agriculture. The Organizing Committee, consisting of J.A. Saunders (Chair), J.M. Gillespie, L. Kosak-Channing, E.H. Lee, J.P. Mack, B.F. Matthews, M.M. Millard and G. Wolfhard, is indebted to the University of Maryland for hosting this meeting.

James A. Saunders
Lynn Kosak-Channing
Eric E. Conn

April, 1987

CONTENTS

CONTENTS

Chapter One

SULFUR DIOXIDE AND CHLOROPLAST METABOLISM

RUTH ALSCHER, MICHAEL FRANZ AND C.W. JESKE

Boyce Thompson Institute
Tower Road
Ithaca, New York 14853

INTRODUCTION

The study of metabolic interactions between SO_2 and
plants is a complex one due to the dual role of sulfur as
a toxic element (as SO_2) and as an essential nutrient in
plant metabolism. Plant cells possess pathways for sulfur
assimilation and a pathway for the removal of excess
sulfur from the plant.[1,2] The toxic effects of SO_2 can be
described as "abnormal" or pathological events which occur
when the plant/cell/organelle is "overloaded" with reactive

(oxidizing) sulfur species of exogenous origin which
exert their damaging influence before they can be
metabolized and/or detoxified. An alternative and equally
important description would involve the "normal" processes
which are adversely affected by the sulfur species. Lastly,
but not least, metabolic responses to SO_2 damage can
usefully be approached through a study of the mechanisms
which confer resistance to the toxic species on the
plant/cell/organelle.

In order to understand the processes through which
the plant cell is damaged by sulfur species, three
distinctly different types of information must be gathered.

First, the pathology of SO_2 damage must be understood
and its possible relationship to the metabolic fate of
sulfur of exogenous origin determined.

Secondly, the pathways, processes, and molecules whose
functioning is altered by the presence of sulfur species
must be identified. Comparisons of the relative susceptibilities of these various functions in closely related
species or cultivars must be included in order to "rank"
the functions in terms of their potential contribution to
the responses of the whole plant or cell to SO_2.

Thirdly, data must be accumulated which relate
metabolic resistance mechanisms to differential susceptibilities of cultivars or species to SO_2.

The purpose of this essay is to assemble a picture
of what is known concerning the events associated with the
interaction of SO_2 with photosynthetic tissue or cells.
To do this, the various types of data listed above must
be integrated and evaluated as an interacting system.

EFFECTS OF SULFUR DIOXIDE ON CHLOROPLAST METABOLISM

Thylakoid Function - A Site of Sulfur Dioxide Action
Associated with Photosystem II (PS II)

The data of Shimazaki and Sugahara, 1979, 1980[3,4] and
Shimazaki et al., 1984[5] suggest a site of SO_2 action
associated with Photosystem II. They determined the

effects of exposure to SO_2 at a high level (1 or 2 ppm)
on the electron-transport capacities of thylakoids isolated
from the leaves of intact lettuce and spinach plants.
Whole-chain electron-transport (water to $NADP^+$) was affected
by this treatment. The site of action was further identified by the use of dichlorophenol indophenol (DCIP) as an
electron acceptor in place of $NADP^+$, indicating an
association with PS II. The time course of this inhibition
showed that exposure to 2 ppm results in a more rapid
inhibition (60% after 5 hours) as compared with 1 ppm
(20% inhibition after 6 hours). Photophosphorylation was
also affected by this type of exposure.

Leaves of plants which had been subjected to the
exposure regime showed decreased photosynthesis 24 hours
after they had been removed from the polluted environment
indicating irreversible damage to the photosynthetic
apparatus.

It appears that the inhibitory effect of SO_2 is
exerted only in the light, since sodium sulfite treatment
(2.5 mM) of spinach disks in the dark had no effect on
subsequent rates of PS II electron transport while rates
from disks illuminated during the sulfite exposure dropped
to less than 10% of control values. Low temperature
fluorescence emission spectra of thylakoids isolated from
the treated disks provided further evidence for a site of
injury associated with PS II.

The results of Libera et al. (1973)[6] and those of
Cerovic et al. (1982)[7] provide a context in which to
evaluate the results of the Sugahara group and ultimately
to assess their relevance to overall effects of SO_2 in the
atmosphere on the photosynthetic process. Libera et al.
demonstrated that exposure of spinach chloroplasts to
sulfite concentrations up to 3 mM resulted in stimulation
of linear, ADP-stimulated electron transport. Stimulation
was as high as 40% at the 1 mM level under their conditions.
Using the known inhibition of PS II activity by Tris buffer
as a probe, they demonstrated that sulfite at and below
1 mM was acting as an electron donor to PS II, since
addition of sulfite resulted in restoration of electron
transport activities previously impaired by exposure to
Tris. Thus, it appears that SO_2/sulfite interacts with
PS II but the nature of the interaction is concentration
dependent.

SULFUR DIOXIDE AND PHOTOPHOSPHORYLATION

Cerovic et al. (1982)[7] and Wellburn (1984)[8] each
provide further evidence for an effect of SO_2/sulfite on
photophosphorylation. Wellburn demonstrated a decrease in
the quenching of the fluorescence of 9-amino-acredine in
oat thylakoids which had been exposed to 1 mM sulfite.
This result consitutes evidence that exposure to sulfite
under these conditions results in a decreased capacity to
form the transthylakoid proton gradient required for
photophosphorylation. Cerovic et al. (1982)[7] demonstrated
that bubbling SO_2 through a pea thylakoid suspension
decreased rates of ATP formation, but did not affect rates
of linear, ADP-stimulated electron transport. The
inhibition was entirely reversed by a simple dilution of
the sulfite-treated thylakoids with additional resuspension
buffer. It appears that photophosphorylation is more
sensitive to sulfite than is electron transport itself.

RANKING CHLOROPLAST FUNCTIONS FOR SULFITE SENSITIVITY

Libera et al. (1973)[6] also studied the effect of
exposure to sulfite at varying concentrations on the
carbon dioxide fixation capacities of isolated intact
spinach chloroplasts. Exposure to levels of sulfite
below 1 mM resulted in a stimulation of carbon fixation,
with 250 µM producing an increase of up to 75%. However,
the 5 mM level which stimulated electron transport at
PS II inhibited carbon fixation. Thus, a site or sites
within the photosynthetic machinery must be more sensitive
to the presence of SO_2/sulfite than is either electron
transport or photophosphorylation.

Results obtained in our laboratory constitute
further evidence for the existence of metabolic sites
within the chloroplast which are more sensitive to sulfite
than are thylakoid functions. Carbon fixation in the
protoplasts of two cultivars of pea proved to be differentially sensitive to the presence of 1.5 mM sulfite as
is shown in Figure 1a. However, rates of uncoupled
electron transport in thylakoids isolated from these
protoplasts were slightly, but equally, affected, as
shown in Figure 1b. Qualitatively similar results were
also obtained for rates of ADP-stimulated electron
transport.

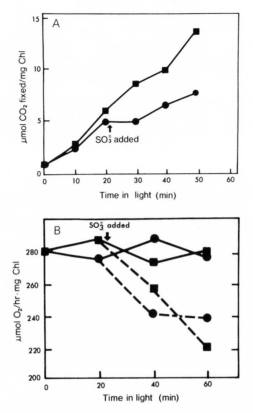

Fig. 1a. Time course for the effect of 1.5 mM $SO_3^=$ on CO_2 fixation in protoplasts from cv "Progress" (■) and cv "Nugget" (●). $SO_3^=$ was added to the incubation medium after 20 minutes of illumination. Results presented are the mean of three experiments performed on three separate occasions. Control values for CO_2 fixation (no sulfite added): cv "Progress", 20.5 ± 1.8 μmol carbon mg chl^{-1} hr^{-1}, (n=2); cv "Nugget", 23.5 ± 1.8 μmol carbon mg chl^{-1} hr^{-1}, (n=2).

Fig. 1b. The effect of NH_4Cl uncoupled electron transfer in chloroplasts isolated from protoplasts of cv "Progress" (■) and cv "Nugget" (●) exposed to 1.5 mM $SO_3^=$. ——— = control; ---- = sulfite treated. Control values (after 20 mins of illumination): cv "Progress", 276.2 ± 5.68 μmol O_2 mg $chl^{-1}hr^{-1}$; cv "Nugget", 287.8 ± 5.68 μmol O_2 mg chl^{-1} hr^{-1}.

It is not surprising to find that sites more sensitive to sulfite/SO_2 exist than those identified by exposures at 2 ppm, since 2 ppm is an order of magnitude higher than the levels of SO_2 reported to inhibit apparent photosynthesis in some species. Black,[9] for example, has compiled results from a number of sources showing a range of sensitivities of apparent photosynthesis to SO_2 from less than 0.1 ppm for pea to 0.8 ppm for Atriplex. It is important to realize, however, that such comparisons (i.e., 2 ppm versus the lower range of concentrations) must be regarded only as guidelines, and have the potential to be misleading since the response of plants to atmospheric pollution is a function of many additional environmental and developmental factors.

Sulfur Dioxide and Stromal Enzymes

Exposures to levels of SO_2 that do not result in irreversible damage sometimes cause a temporary inhibition of net photosynthesis, with recovery to control rates after removal of the plants from the SO_2 atmosphere, e.g., soybeans exposed to 0.5 ppm for one hour.[10] Thus, SO_2 can, under certain circumstances, reversibly disrupt the operation of the photosynthetic machinery. One possible target for a reversible disruption such as this is a regulatory function associated with chloroplast metabolism which is affected by SO_2.

A body of evidence has accumulated over the past 15 years which demonstrates the existence of a light-controlled, sulfhydryl group mediated, regulation system for carbon fixation called light "modulation", since it acts to activate and to inactivate regulatory chloroplast enzymes depending on prevailing conditions.[11-13] Sulfur dioxide, by virtue of its chemical properties, is well-suited to interact with the light modulation system, and some evidence has accumulated which suggests that this is indeed the case.

One possible consequence of sulfite oxidation is the production of H_2O_2. H_2O_2 has been shown to inhibit photosynthetic carbon fixation at the level of the bisphosphatase enzymes by causing an oxidation of essential thiol groups, resulting in a reversion of these regulatory enzymes to their inactive, oxidized forms.[14] Thus, a consequence of sulfite oxidation in illuminated chloroplasts could be the inactivation of the light-regulated enzymes of the Calvin cycle and hence the inhibition of photosynthesis. Direct evidence for this has been provided by the results of Tanaka et al.[15,16] who demonstrated that H_2O_2 accumulates in chloroplasts isolated from spinach leaves fumigated with SO_2.

Several light-modulated, sulfhydryl-containing chloroplast enzymes, including fructose-1,6-bisphosphatase, were inhibited under these exposure conditions, whereas enzymes which are not regulated by the thioredoxin-ferredoxin pathway for reductive activation (i.e., ribulose-1,5-bisphosphate carboxylase oxygenase (Rubisco)) were much less affected. The results on enzyme activity were substantiated and related to photosynthetic activity through various additional measurements. The decrease in enzyme activities which occurred at the onset of fumigation was correlated in time with a decrease in apparent photosynthesis. Apparent photosynthesis did not recover fully after removal of the plants from the SO_2 atmosphere. However, the sulfhydryl-containing enzymes rapidly and fully regained activity. A decrease in fructose-6-phosphate levels (the product of the reaction catalyzed by fructose-1,6-bisphosphatase) and an increase in fructose-1,6-bisphosphate levels (the substrate) occurred as a result of exposure to SO_2. This result, showing that a condition exists under which light-modulated enzymes are more sensitive to SO_2 than the carboxylase, is an important one, since in a previous report[17] the enzyme was shown to be inhibited by sulfite both competitively and non-competitively ($Ki = 15$ mM) with respect to CO_2. The results of Tanaka et al. allow the carboxylase enzyme to be ranked vis-a-vis light-modulated sulfhydryl-containing enzymes with respect to SO_2 sensitivity. The mechanism of inactivation by SO_2 involves sulfhydryl groups since addition of dithiothreitol (DTT) to chloroplasts isolated from fumigated leaves resulted in a reversal of the inactivation of, in this case, $NADP^+$-dependent glyceraldehyde-3-phosphate dehydrogenase. Results obtained in our laboratory (Table 1) also demonstrate an inactivation of light-modulated enzymes in sulfite-treated protoplasts from the pea cultivars shown in Figure 1a to be sulfite-sensitive with respect to carbon fixation. Under these experimental conditions (2 ppm SO_2, 15 mM sulfite) a correlation exists between the response of photosynthesis to SO_2/sulfite and that of the light modulation system.

Table 1. Effect of exposure to light and $SO_3^=$ on the activities of three light-modulated enzymes.

| Treatment | Time In Light (min) | cv | n | Specific Activity of Enzymes (μ mol/mg chl.hr) | | |
				FbPase	MDH	GAPDH
No addition	0	P	4	32.9±0.85	3.80±0.35	50.7± 1.6
		N		40.2±4.1	8.90±0.8	78.1± 6.3
	20	P	4	50.8±0.8	6.90±0.15	182.4±16.1
		N		55.4±3.4	10.1 ±1.15	204.0± 7.3
1.5 mM $SO_3^=$	40	P	4	36.2±2.9	6.16±0.5	98.2±14.3
		N		38.1±4.1	5.72±0.85	120.5± 4.8
	60	P	4	49.8±1.3	6.49±0.9	72.0± 1.8
		N		27.8±2.0	5.70±0.6	74.1± 2.9
No addition	60	P	2	45.2±0.7	6.39±0.1	163.8±11.4
		N		43.1±1.1	7.87±0.1	167.8±12.2

Samples were taken for measurements of enzyme activity from protoplasts of pea, cv "Progress" (P) or "Nugget" (N). Values shown are means ± standard error; n = number of separate experiments; FbPase = fructose-1,6-bisphosphatase; MDH = malic dehydrogenase; GAPDH = glyceraldehyde-3-phosphate dehydrogenase.

However, in an in vivo study in our laboratory (Alscher et al.[60]) we found that exposure of intact pea plants to 0.8 ppm SO_2 resulted in a temporary inhibition of photosynthesis which was not reflected in an inactivation of light-modulated enzymes. (Neither the response of the stomatal apparatus nor that of electron transport correlated with the response of photosynthesis either.) Photosynthesis can be influenced by SO_2, therefore, without influencing light-modulation. We interpret these data as an expression of the response of another, more sensitive, site of action of SO_2 within the chloroplast.

Sulfite itself has been shown to have direct effects on regulatory enzymes of carbon metabolism. Ziegler's group has demonstrated that sulfite inhibits light activation of NADP$^+$-dependent glyceraldehyde-3-phosphate dehydrogenase[18] and malic dehydrogenase.[19]

Since sulfite can react directly with the sulfhydryl-containing enzymes of the chloroplast, results obtained in our laboratory[20] and elsewhere[21] demonstrating an inhibition of light modulation in broken chloroplasts by sulfite and SO_2 respectively, could be either an expression of a direct action of sulfite or of the inhibitory effect of H_2O_2 produced as a consequence of the presence of sulfite. This same interpretation could be applied to the results of Tanaka et al.[15,16] but may be unlikely, at least at lower levels of SO_2, for reasons discussed below.

Fumigation of spinach plants with levels of SO_2 much lower than those used in the studies reported above (0.028 ppm versus 0.8, 2 and 63 ppm), on the other hand, has been shown to stimulate the light activation of chloroplast enzymes; this activation appears to take place through generation of thylakoid-association thiol groups.[22] Since Ziegler and Libera[23] and Libera et al.[6] found that concentrations of sulfite below 1 mM stimulated carbon fixation and electron transport in isolated chloroplasts, an enhancement of the light-activation process at low levels of SO_2 is not surprising. If more reductant is available for the needs of carbon fixation, there is also more reductant to bring about the reductive activation of regulatory chloroplast enzymes. Interestingly, increases in apparent photosynthesis elicited by exposure to low levels of SO_2 have also been reported.[9]

On the basis of the available evidence, the various chloroplast functions known to be affected by SO_2 can be ranked as shown in Table 2.

THE METABOLIC FATE OF SULFITE OF EXOGENOUS ORIGIN

Within the leaf cell, SO_2 in its aqueous form as sulfite[24] has been shown to accumulate in the chloroplast. Once in the chloroplast, sulfite can either be oxidized[25] or reduced.[1,2] Sulfate is the principal oxidation product[24] and it accumulates to high levels in the tissues

Table 2.   Chloroplast metabolism and sulfur dioxide.

Relative Susceptibilities

Rubisco

Least Sensitive

Photosystem II

Proton Gradient Formation-
photophosphorylation

Light-modulated enzymes

?   Most Sensitive

---

of plants which have been exposed to SO_2 at sublethal doses.

The mechanism by which sulfite is oxidized to sulfate in the chloroplast has been described by Asada.[26] It is an aerobic process which is initiated by light and is mediated by the photosynthetic electron transport chain. Oxygen is consumed in the process and the superoxide radical is formed. The ratio of the oxidation rate of sulfite to the rate of electron transfer was such that the reaction sequence shown below ((1)-(7)) was proposed as the mechanism of sulfite photo-oxidation, initiated by superoxide in the chloroplast.

$$SO_3^{2-} + O_2^- + 3\ H^+ \rightarrow HSO_3\cdot + 2\ OH\cdot \quad (1)$$

$$SO_3^{2-} + OH\cdot + 2\ H^+ \rightarrow HSO_3\cdot + H_2O \quad (2) \quad \text{chain propagation}$$

$$HSO_3\cdot + O_2 \rightarrow SO_3 + O_2^- + H^+ \quad (3)$$

$$HSO_3\cdot + OH\cdot \rightarrow SO_3 + H_2O \quad (4)$$

$$2\ HSO_3\cdot \rightarrow SO_3 + SO_3^{2-} + 2\ H^+ \quad (5) \quad \text{chain termination}$$

$$SO_3 + H_2O \rightarrow SO_4^{2-} + 2\ H^+ \quad (6) \quad \text{hydration of SO}$$

$$2\ OH\cdot \rightarrow H_2O_2 \quad (7)$$

Other oxidizing free radicals such as OH· are formed in the course of the chain reaction. (Evidence for the participation of free radicals in the process was provided by the results of Asada and Kiso,[25] showing that free-radical scavengers such as mannitol and myo-inositol inhibited the photo-oxidation of sulfite by spinach chloroplasts.) Sulfate is formed, finally, following chain termination through hydration of the SO_3 molecule. Superoxide dismutase which acts on superoxide with the resultant production of hydrogen peroxide inhibits the oxidation of sulfite by illuminated chloroplasts.[26] This result is to be expected if superoxide acts as a chain initiator and carrier as is shown in reactions (1)-(7).

Sulfite reduction instead of oxidation can also occur as a result of electron transport. An enzyme with sulfite reductase activity has been described which utilizes reduced ferredoxin as a source of reductant.[27] Rothermel and Alscher[28] demonstrated that sulfite was metabolized in cucumber cells in a light-dependent manner apparently utilizing reduced ferredoxin as a source of reductant.

The relative importance of these two pathways for the metabolism of sulfite is unclear although the oxidation route carries with it the potential danger of free radical generation. In addition to this unresolved question, little information is available concerning the levels of sulfite which accumulate in the chloroplasts of plants which have been exposed to SO_2, or in sulfite-treated cells. Since the metabolic consequences for the chloroplast in the presence of sulfite of exogenous origin are several, and differentially concentration-dependent, this is an unfortunate gap in our understanding of the mechanism or mechanisms affected by exposure to any given concentration of SO_2. Without this information, it is not possible to assess the wealth of in vivo and in vitro data which show effects of SO_2 or sulfite on photosynthesis and other metabolic processes.

Whichever pathway is taken, be it oxidation or reduction, the speed of detoxification of SO_2 could be of importance in the context of plant resistance to the pollutant. Miller and Xerikos,[29] using a series of soybean cultivars, found a strong correlation between "residence time" of sulfite with resistance to SO_2, ranked according to degree of foliar injury. In our laboratory we adapted

Table 3. Effect of the presence of 1 ppm sulfur dioxide on
sulfite concentrations "in vivo" in soybean cv. 'Hark' and
'Beeson' after exposure to light.

| | Cultivar | \multicolumn{4}{c}{Minutes Light} |
| | | 30 | | 60 | |

| | | \multicolumn{4}{c}{µM Sulfite "in vivo"} | | | |

	Cultivar	Control	SO_2	Control	SO_2
A.	'Hark'	350	500	60	1,110
	'Beeson'	410	120	540	70
B.	'Hark'			250	880
	'Beeson'			416	760

All values shown were obtained from four separate exposures.
For each exposure, duplicate or triplicate leaf samples
were taken and sulfite levels determined on leaf homogenates. Plants were grown in either the field (A) or
greenhouse (B). In the case of the field exposures, leaf
samples were taken 30 and 60 min after sunrise. The light
intensity at that time was ca 150 µE $m^{-2}s^{-1}$. Light
intensities in the controlled environment chamber were
also 150 µE $m^{-2}s^{-1}$. Sulfite levels were determined by the
method of Bourbon et al. (1971), Pollution Atmospherique
52: 271-275.

a method used for the determination of SO_2 levels in the
air for use in the determination of sulfite levels
attained in fumigated leaves of two soybean cultivars
whose SO_2 susceptibilities in the field with respect to
yield are known.[30] The results are shown in Table 3. A
higher levels of sulfite accumulated in the leaves of the
more sensitive cultivar in both field and controlled
environment exposures. The results are expressed as
sulfite concentrations "in vivo", which we defined
arbitrarily as the concentration of sulfite expressed as

micromoles of sulfite per liter cell water. The underlying assumption that the interior of the leaf is a
homogeneous aqueous suspension is obviously a convention,
at best! These values represent, perhaps, a lower limit
for sulfite concentration in vivo. The values which were
obtained suggest that exposure of soybean plants (R7
developmental stage[31]) to 1 ppm SO_2 results in the
accumulation of sulfite concentrations approaching 1
millimolar, although the levels may be considerably higher
at sites of action within the cell. The correspondence
between 1 ppm SO_2 and 1 mM sulfite may be useful as an
approximate indicator, allowing preliminary judgments to
be made concerning possible metabolic sites affected by
any given dose of sulfur dioxide.

Sulfite accumulation was also greater in the cultivar
more sensitive to SO_2 in the case of the two pea cultivars
described previously (Franz and Alscher, unpublished;
Alscher et al.[60]), whether the exposure was carried out
on whole plants (0.8 ppm SO_2) or protoplasts (1.5 mM
sulfite). In these cases, we determined sulfite levels by
the method described in Rothermel and Alscher[28] on leaf
homogenates obtained from exposed plants and on isolated
intact chloroplasts obtained from sulfite-treated protoplasts.

ACTIVE OXYGEN AND SULFUR DIOXIDE TOXICITY

From the data discussed above it seems that the
speed of detoxification of sulfur species of exogenous
origin is an important factor contributing to a plant's
ability to withstand damage from SO_2. This could be
because of the damage which sulfite can exert while
present in any particular subcellular compartment. For
instance, sulfite has the capacity to interact with
disulfide bonds causing disruption of the tertiary
structure, and hence the function, of proteins.[32] The
basis of differential tolerance to SO_2 seen in both
soybean and pea may partially reflect the differential
levels of sulfite which accumulate in the leaf cell as a
result of exposure. Therefore in the case of the more
resistant cultivar, larger doses of SO_2 would be needed to
produce the equivalent sulfite concentration in vivo. This
holds true, of course, whatever the mechanism of SO_2 damage
proves to be.

Evidence has accumulated, however, to suggest that there exists at least one other major mechanism by which SO_2 damages the chloroplast and possibly the plant cell as a whole. Tanaka et al.[15,16] demonstrated that there is an oxygen requirement for the inactivation of chloroplast glyceraldehyde-3-phosphate dehydrogenase by 2 ppm SO_2. If this were an expression of a simple interaction of sulfite with disulfide bridges on the enzyme, oxygen would not be required for the reaction. Tanaka et al. demonstrated also that under their fumigation conditions the ratio of NADPH to $NADP^+$ increases over the course of the fumigation. This increase will occur when CO_2 fixation is inhibited but reductant continues to be produced as a consequence of photosynthetic electron transport. Under these conditions, there is no $NADP^+$ available to be reduced by ferredoxin, and molecular oxygen is reduced instead, as is shown in Figure 2. The univalent reduction of molecular oxygen results in the production of the superoxide radical which, as described above, participates in the photo-oxidation of sulfite to sulfate.

This reduction of molecular oxygen (named the Mehler reaction after its discoverer) is, however, also a 'normal' event in chloroplast metabolism. It is thought that pseudocyclic electron transport, as it is called, acts as a "safety valve" for the photosynthetic electron transport chain to prevent the over-reduction of carriers, and also to provide the extra ATP needed for carbon fixation when linear and cyclic electron transport do not provide sufficient amounts.[33] Superoxide can also be produced by direct reduction from the PS I reaction center as is shown in Figure 2. The great abundance of oxygen in illuminated active chloroplasts makes this mechanism readily available to serve its safety valve function as needed.

The consequences of the "normality" of superoxide production for SO_2 oxidation and toxicity are several. Firstly, superoxide will always be present in a photosynthetically active chloroplast to initiate the chain reac reaction associated with sulfite oxidation. Secondly, since superoxide production is not an extraordinary event, pathways for its removal must exist within the chloroplast. Thirdly, other free radicals which are formed as a result of exogenous sulfite photo-oxidation (not a physiological

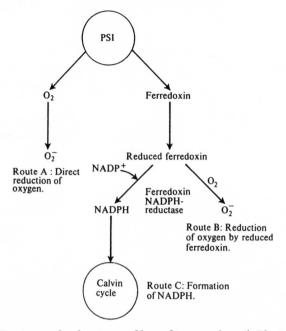

Fig. 2. Routes of electron flow from reduced Photosystem
I. Reduced ferredoxin reduces oxygen to O_2^- as well as
passing electrons onto $NADP^+$. The electron acceptor of
PS I itself slowly reduces oxygen to O_2^-. Courtesy of
Oxford University Press.

event since sulfite of endogenous origin is assimilated
quite differently[34]), are likely to be more toxic to the
chloroplast/cell than the superoxide radical itself.
Lastly, and a great deal more hypothetically, the toxicity
of SO_2 derived species is likely to be a consequence of
overriding of resistance mechanisms rather than a unique
and pathological metabolic event. Data supporting these
proposals are presented below.

The Metabolism of Superoxide Anion in the Chloroplast

Superoxide anion is converted to hydrogen peroxide in
the chloroplast through the action of superoxide dismutase
(SOD). SOD is ubiquitous in aerobic organisms. It seems
to serve the same purpose throughout nature since it is

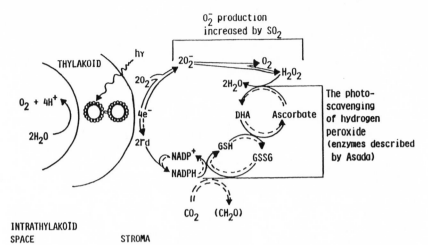

Fig. 3.   Formation and scavenging of superoxide and
hydrogen peroxide in chloroplasts.   Courtesy of Japanese
Society of Plant Physiologists.

quite specific for its substrate.[35]   Photosynthetic plant
cells contain at least three electrophoretically distinct
forms of the enzyme.[36]   Two of these are cyanide-sensitive
and contain copper and zinc.   The third is cyanide-
insensitive and contains manganese in the place of copper
and zinc.   Mn-SOD activity is described as usually being
associated with the mitochondrion, while one of the Cu-Zn
enzymes is known to be associated with the chloroplast[37]
and the other with the cytosol.   All three forms are
genetically distinct and coded for in the nucleus.[38]   The
unique chloroplastic enzyme was thought to be located
primarily in the stroma with about 30% being membrane-
bound.[39]   The elegant studies of Asada and his co-workers,
however, have expanded our understanding of chloroplast
SOD and its functioning.   The hydrogen peroxide produced
as a result of SOD activity is itself toxic (i.e., it can
inactivate light-modulated enzymes) and must be removed
rapidly as a consequence.   This is achieved through a
metabolic cycle involving successive oxidation and re-
reduction of glutathione, ascorbic acid and NADPH (Fig.
3).[40-42]   Ascorbate can also react with superoxide.[35]
In light the hydrogen peroxide formed is rapidly scavenged
through the action of this cycle.[43]   Takahashi and Asada[44]

showed that although the permeability of thylakoids to the superoxide radical is very low, a potentially dangerous situation for the thylakoid membrane exists if unreacted superoxide diffuses into the acidic thylakoid, or is produced there (through damage to lumen-exposed electron transport chain components). In this case a highly reactive form of superoxide, HO_2, is produced. Hayakawa et al.[45,46] demonstrated, however, that there is another SOD within the lumen of the thylakoid which is immuno-logically distinct from the previously described molecules, and is apparently a third type of SOD associated with the chloroplast. This enzyme is of the cyanide-insensitive, Mn-containing group. SOD is located at each side of the chloroplast membrane, therefore, as well as being bound to it.

It appears that the other components of the photo-scavenging cycle are present only in the stroma where they occur at very high concentrations. Glutathione occurs in the 1 to 5 mM range and ascorbate in the 25 mM range.[33] Glutathione exists primarily in the reduced form in the chloroplast, both in the light and in the dark, due to the action of glutathione reductase.[47] The oxidized form of glutathione (GSSG) can inactivate a number of enzymes, probably by forming mixed disulfides with them.[35] Since hydrogen peroxide at levels as low as 10 μM inhibits carbon fixation by 50 percent[48] it is not surprising to find that components of the pathway for its removal are present in such abundance in the chloroplast.

Superoxide and Related Species: Mechanisms of Damage

Although it is not as reactive in aqueous solution, superoxide is a powerful nucleophile and base in organic solvents. It can react with protons to form the hydro-peroxyl radical or it can act as a reducing agent with thylakoid components such as plastocyanin.[35] In the environment of the membrane, it can also react with hydrogen peroxide forming the hydroxyl radical (Reaction 8), a species which can destroy phospholipids through a nucleophilic attack on the carbonyl groups of the ester moieties, releasing the free fatty acids from the glycerol portion of the molecule.

$$H_2O_2 + O_2^- \rightarrow O_2 + OH^- + OH\cdot \qquad (8)$$

The reaction between superoxide and hydrogen peroxide can
only take place in the presence of transition metal ions.
Copper and iron, both present in thylakoid components,
can serve this function.

If the production of hydroxyl radical were a pathway
through which superoxide anion damaged membranes, then
scavengers of the hydroxyl radical such as mannitol should
protect against damage exerted by superoxide generating
species. Asada[26] demonstrated that mannitol did indeed
inhibit sulfite photo-oxidation in spinach chloroplasts.
Lizada and Yang[49] found that the aerobic oxidation of
sulfite resulted in the peroxidation of linoleic acid
through the formation of conjugated dienes and, eventually,
lipid hydroperoxides. The peroxidation of polyunsaturated
membrane lipids will lead to changes in permeability
properties and in fluidity, causing the membrane
"leakiness" so commonly observed in tissues which have
been subjected to oxidative stress. No effect of hydroxyl
scavengers on sulfite-induced linoleic acid peroxidation
was detected in Lizada and Yang's system, although other
free radical scavengers such as alpha-tocopherol (a major
thylakoid constituent) were very effective in inhibiting
the process. The basis for the discrepancy between these
two reports is not clear. It is possible that the Lizada
and Yang system did not contain the metal ions which are
necessary to catalyze the hydroxyl radical formation
reaction. However, there is agreement that the photo-
oxidation of sulfite involves the production of free
radicals and that this can result in lipid peroxidation.

OXYGEN TOXICITY AND SULFUR DIOXIDE: PROTECTIVE MECHANISMS

The Induction of Superoxide Dismutase by Exposure to
Sulfur Dioxide

Given the distribution of SOD within the chloroplast,
it is plausible to propose that it serves a protective
function for the chloroplast in the highly aerobic environment which occurs when photosynthesis is taking place.
Tanaka and Sugahara[50] showed that poplar plants which had
been exposed to a relatively low level of SO_2 (0.1 ppm)
were more resistant than control plants to subsequent
damage exerted by exposure to 2 ppm SO_2 (Figs. 4 and 5).
Rabinowitch and Fridovich[51] showed that Cu-Zn SOD was

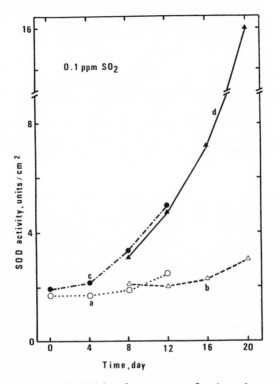

Fig. 4   Induction of SOD by long term fumigation with 0.1 ppm SO_2. The poplar plant at 30 days after cutting was fumigated with 0.1 ppm SO_2 for 20 days. SOD activities of leaves in the fifth to eighth positions at eight days after 0.1 ppm SO_2-fumigation were also followed. Curves a, control, older leaves; b, control, leaves developing during fumigation; c, 0.1 ppm SO_2, older leaves; d, 0.1 ppm SO_2, leaves developing during fumigation. Courtesy of National Institute of Environmental Studies.

induced in Chlorella cells which had been exposed to sulfite and that these cells were subsequently more resistant to the herbicide Paraquat than uninduced cells. Thus, the plant cell possesses at least one system which protects it against two different oxidative stresses and which can respond to that stress with a metabolic reinforcement, as it were.

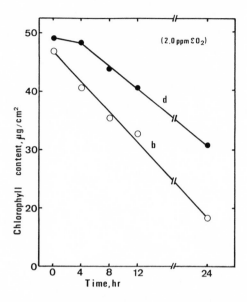

Fig. 5. Effect of fumigation with 2.0 ppm SO_2 on poplar plants in which the content of SOD was increased with 0.1 ppm SO_2-fumination. Both poplar leaves were fumigated with 2.0 ppm SO_2. Curves b, control; d, 2.0 ppm SO_2 twenty days after fumigation with 0.1 ppm SO_2. Courtesy of National Institute of Environmental Studies.

The Role of Glutathione and Ascorbate

Glutathione acts to remove hydrogen peroxide generated as a result of SOD activity (Fig. 3). It can also scavenge hydroxyl radicals itself.[35] Glutathione reductase acts to keep the large (1-5 mM) pool of glutathione predominantly (> 90%) in the reduced form. Under conditions of oxidative stress such as exposure to SO_2, ozone, low temperatures or drought, the level of reduced glutathione in foliar tissue has been shown to increase above control levels.[52-57]

Thus, glutathione appears to be part of a resistance mechanism which is not specific to SO_2. Rather, the chloroplast appears to possess a mechanism for responding to the production of toxic species which is applicable to many different oxidative stresses. Since under unstressed conditions the majority of glutathione exists in the

reduced form, it is possible that this increase is an
expression of an induction process, as appears to be the
case for SOD. Alternatively, it may be the expression of
the activation of a rate-limiting enzyme in the pathway
for glutathione production. Information is not available
concerning stress-induced changes in reduced glutathione
concentrations in specific subcellular compartments.
Thus, it is not possible to determine if this response
to oxidative stress is due primarily to the chloroplast
photoscavenging system. Given the high probability of
radical species formation, however, the stress-induced
production of antioxidant in the chloroplast is highly
plausible.

Ascorbate is an important component in the photo-
scavenging cycle (Fig. 3) and in addition can act as a
scavenger of hydroxyl radicals itself.[35] Beck et al.[58]
have shown that ascorbate is transported across the
chloroplast envelope by a specific carrier system. It
appears that the ascorbate used in the scavening reactions
of the chloroplast is produced in the cytosol and can be
imported across the chloroplast envelope as needed. This
import mechanism or the production pathway for ascorbate
in the cytosol might be expected, therefore, to be
responsive to oxidative stress.

CONCLUSION

Any disruption of chloroplast metabolism which leads
to an inhibition of carbon fixation or electron transport
will bring about an increase in toxic oxygen species.
Depending on the dose, sulfur dioxide/sulfite treatment
can lead to such disruptions in stromal or thylakoid
function. The SOD-ascorbate-glutathione pathway is a
major protection mechanism against oxygen toxicity (Tables
4 and 5). A crucial aspect of the mechanism is the
inducibility of the components of the pathway on exposure
to oxidative stress. Only when protection systems have
been overwhelmed, will toxic species persist long enough
to damage essential processes and macromolecules. Only
under these conditions should the plant cell or organelle
be damaged.

Boyer[59] reports that inability to withstand environmental stress is the primary source of agronomic loss in

Table 4. Chloroplast constituents which protect against oxygen toxicity.

Molecule	Location	Response to Metabolic Stress	Reference
SOD	Thylakoid (Mn)		
	Stroma (Cu-Zn)	Induction of enzyme	50,51
	Thylakoid lumen (Cu-Zn)	Dismutation of superoxide	45,26
Glutathione	Stroma	Increase in GSH	56
		Scavenge OH· radicals ⎤	35,33
		Scavenge singlet oxygen ⎦	
		Enzymatically reduced DHA	40,41
Ascorbate	Stroma	Import from cytosol?	58
		Scavenge OH· radicals ⎤	35,33
		Scavenge singlet oxygen ⎦	
		Enzymatically reduced H_2O_2	40,41
Alpha-tocopherol	Thylakoid	Scavenge peroxy, alkoxy ⎤ Radicals ⎦	35
Carotenoids	Thylakoid	Quench singlet oxygen	35,33

Table 5. Toxic species formed as a result of oxidative stress.

Reactive Species	Formation	Metabolic Consequence
Superoxide	Mehler reaction Sulfite oxidation	Formation of reactive species
Singlet oxygen	Transfer from excited CHL	Lipid peroxidation
Hydroxyl radical	$H_2O_2 + O_2^-$	Reaction with all major classes of macromolecules
Hydrogen peroxide	SOD activity	Inactivation of SH-enzymes
GSSG	Oxidation of GSH	Enzyme inactivation
Peroxy radical	Lipid peroxidation	Change in membrane properties

Any change in membrane properties can be expected to result in changes in its function, e.g., lowered electron transport in the case of the thylakoid.

U.S. agriculture. One object of future work in the area of environmental biology could well be a better understanding of stress resistance mechanisms. Aided by the tools of molecular biology and genetics, we can hope to increase productivity through an enhancement of the ability of plant cells to withstand the oxidative stresses which are the norm under many environmental conditions.

REFERENCES

1. RENNENBERG, H. 1982. Glutathione metabolism and possible biological roles in higher plant. Phytochemistry 21: 2771-2781.

2. FILNER, P., H. RENNENBERG, J. SEKIYA, R.A. BRESSAN,
    L.G. WILSON. 1984. Biosynthesis and emission of
    hydrogen sulfide by higher plants. In Gaseous Air
    Pollutants and Plant Metabolism. (M.J. Koziol,
    F.R. Whatley, eds.), Butterworths, London, pp.
    291-312.
3. SHIMAZAKI, K., K. SUGAHARA. 1979. Specific inhibition
    of photosystem II activity in chloroplasts by
    fumigation of spinach leaves with SO_2. Plant
    Cell Physiol. 20: 947-955.
4. SHIMAZAKI, K., K. SUGAHARA. 1980. Inhibition site
    of the electron transport system in lettuce
    chloroplasts by fumigation of leaves with SO_2.
    Plant Cell Physiol. 21: 125-135.
5. SHIMAZAKI, K., K. NAKAMACHI, N. KONDO, K. SUGAHARA.
    1984. Sulfite inhibition of photosystem II in
    illuminated spinach leaves. Plant Cell Physiol.
    25: 337-341.
6. LIBERA, W., H. ZIEGLER, I. ZIEGLER. 1973. Forderung
    der Hill-reaktion und der CO_2-fixierung in
    isolierten Spinatchloroplasten durch neidere
    sulfitkonzentrationen. Planta 109: 269-279.
7. CEROVIC, Z.G., R. KALEZIC, M. PLESNICAR. 1982. The
    role of photophosphorylation in SO_2 and SO_3^{-2}
    inhibition of photosynthesis in isolated chloro-
    plasts. Planta 156: 249-254.
8. WELLBURN, A.R. 1984. The influence of atmospheric
    pollutants and their cellular products upon
    photophosphorylation and related events. In M.J.
    Koziol, F.R. Whatley, eds., op. cit. Reference 2,
    pp. 203-222.
9. BLACK, V.J. 1982. Effects of sulfur dioxide on
    physiological processes in plants. In Effects of
    Gaseous Air Pollution in Agriculture and Horti-
    culture. (M.H. Unsworth, D.P. Ormrod, eds.),
    Butterworths, London, pp. 67-92.
10. McLAUGHLIN, S.B., D.S. SHRINER, R.K. McCONATHY, L.K.
    MANN. 1979. The effects of SO_2 dosage kinetics
    and exposure frequency on photosynthesis and
    transpiration of kidney beans (Phaseolus vulgaris
    L.). Environ. Exp. Bot. 19: 174-191.
11. BUCHANAN, B.B. 1980. Role of light in the regulation
    of chloroplast enzymes. Annu. Rev. Plant Physiol.
    31: 341-433.
12. BUCHANAN, B.B. 1984. Enzyme photoregulation by the
    ferredoxin/thioredoxin system: Review and update.

Workshop on light-dark modulation of plant enzymes, Wallenfels, Aug. 7-9, 1983, University of Bayreuth.

13. ALSCHER, R. 1983. Effects of SO light-modulated enzyme reactions. In Gaseous Air Pollutants and Plant Metabolism. (M. Koziol, F. Whatley, eds.), Butterworths, London, pp. 181-200.

14. CHARLES, S.A., B. HALLIWELL. 1980. Effect of hydrogen peroxide on spinach (Spinacia oleracea) chloroplast fructose bisphosphatase. Biochem. J. 189: 373-376.

15. TANAKA, K., N. KONDO, K. SUGAHARA. 1982a. Accumulation of hydrogen peroxide in chloroplasts of SO_2-fumigated spinach leaves. Plant Cell Physiol. 23: 999-1007.

16. TANAKA, K., H. MITSUHASHI, N. KONDO, K. SUGAHARA. 1982b. Further evidence for inactivation of fructose-1,6-bisphosphatase at the beginning of SO_2 fumigation. Increase in fructose-1,6-bisphosphate and decrease in fructose-6-phosphate in SO_2-fumigated spinach leaves. Plant Cell Physiol. 23: 1467-1470.

17. ZIEGLER, I. 1972. The effect of SO_3^{2-} on the activity of ribulose-1,5-bisphosphate carboxylase in isolated spinach chloroplast. Planta 103: 155-163.

18. ZIEGLER, I., A. MAREWA, E. SCHOEPE. 1976. Action of sulfite on the substrate kinetics of chloroplastic NADP-dependent glyceraldehyde-3-phosphate dehydrogenase. Phytochemistry 15: 1627-1632.

19. ZIEGLER, I. 1974. Action of sulphite on plant malate dehydrogenase. Phytochemistry 13: 2411-2416.

20. ALSCHER-HERMAN, R. 1982. Effect of sulfite on light activation of fructose-1,6-bisphosphatase in two cultivars of soybean. Environ. Pollut. Ser. A 27: 83-96.

21. ANDERSON. L.E., J.X. DUGGAN. 1977. Inhibition of light modulation of chloroplast enzyme activity by sulfite. Oecologia 28: 147-151.

22. MISZALSKI, Z., I. ZIEGLER. 1979. Increase in chloroplastic thiol groups by SO_2 and its effect on light modulation of NADP-dependent glyceraldehyde-3-phosphate dehydrogenase. Planta 145: 383-387.

23. ZIEGLER, I., W. LIBERA. 1975. The enhancement of CO fixation in isolated chloroplasts by low sulfite concentrations and by ascorbate. Z. Naturforsch Teil. C. 30: 634-637.

24. ZIEGLER, I. 1975. The effect of SO_2 pollution on plant metabolism. Residue Rev. 56: 79-105.

25. ASADA, K., K. KISO. 1973. Initiation of aerobic oxidation of sulfite by illuminated spinach chloroplasts. Eur. J. Biochem. 33: 253-257.

26. ASADA, K. 1980. Formation and scavenging of superoxide in chloroplasts, with relation to injury by sulfur dioxide in studies on the effects of air pollutants on plants and mechanisms of phytotoxicity. Res. Rep. Natl. Inst. Environ. Stud. 11: 165-179.

27. SAWHNEY, S.K., D.D. NICHOLAS. 1975. Nitrite, hydroxylamine and sulphite reductases in wheat leaves. Phytochemistry 14: 1499-1503.

28. ROTHERMEL, B., R. ALSCHER. 1985. A light-enhanced metabolism of sulfite in cells of Cucumis sativus L. cotyledons. Planta 166: 105-110.

29. MILLER, J.E., P.B. XERIKOS. 1979. Residence time of sulphite in SO_2 "sensitive" and "tolerant" soybean cultivars. Environ. Pollut. 18: 259-264.

30. AMUNDSON, R.G. 1983. Yield reduction of soybeans due to exposure to sulfur dioxide and nitrogen dioxide in combination. J. Environ. Qual. 12: 454-459.

31. FEHR, W.R., C.E. CAVINESS. 1977. Stages of soybean development. Cooperative Extension Service, Agric. & Home Ec. Exp. Sta., Iowa State Univ. Spec. Report 80: 1-12.

32. MEANS, G.E., R.E. FEENEY. 1971. Chemical Modification in Proteins. Holden-Day, Inc., San Francisco, pp.

33. HALLIWELL, B. 1981. Chloroplast Metabolism. Clarendon Press, Oxford, 259 pp.

34. ANDERSON, J.W. 1981. Light-energy-dependent processes other than CO_2 assimilation. In The Biochemistry of Plants. (M.D. Hatch, N.J. Boardman, eds.), Academic Press, New York, Vol. 8, pp. 473-500.

35. HALLIWELL, B., J.M.C. GUTTERIDGE. 1985. Free radicals in biology and medicine. Clarendon, Oxford, 340 pp.

36. SEVILLA, F., J. LOPEZ-GORGE, L.A. DEL RIO. 1982. Characterization of a manganese superoxide dismutase from the higher plant Pisum sativum. Plant Physiol. 70: 1321-1326.

37. BAUM, J.M., J.M. CHANDLEE, J.G. SCANDALIOS. 1983. Purification and partial characterization of a genetically defined superoxide dismutase (SOD-1)

associated with maize chloroplast. Plant Physiol.
73: 31-35.

38. BAUM, J.A., J.G. SCANDALIOS. 1981. Isolation and
characterization of the cytosolic and mitochondrial
superoxide dismutases of maize. Arch. Biochem.
Biophys. 206: 249-264.

39. JACKSON, C., J. DENCH, A.L. MOORE, B. HALLIWELL, C.H.
FOYER, D.O. HALL. 1978. Subcellular localization
and identification of superoxide dismutase in the
leaves of higher plants. Eur. J. Biochem. 91:
339-344.

40. NAKANO, Y., K. ASADA. 1981. Hydrogen peroxide is
scavenged by ascorbate-specific peroxidase in
spinach chloroplasts. Plant Cell Physiol. 22:
867-880.

41. GRODEN, D., E. BECK. 1979. H_2O_2 destruction by
ascorbate-dependent systems from chloroplasts.
Biochim. Biophys. Acta 546: 426-435.

42. ANDERSON, J.W., C.H. FOYER, D.A. WALKER. 1983.
Light-dependent reduction of dehydroascorbate and
uptake of exogenous ascorbate by spinach chloroplasts. Planta 158: 442-450.

43. HOSSAIN, M.A., K. ASADA. 1984. Inactivation of
ascorbate peroxidase in spinach chloroplasts on
dark addition of hydrogen peroxide: Its protection
by ascorbate. Plant Cell Physiol. 25: 1285-1295.

44. TAKAHASHI, M., K. ASADA. 1983. Superoxide anion
permeability of phospholipid membranes and
chloroplast thylakoids. Arch. Biochem. Biophys.
226: 558-566.

45. HAYAKAWA, T., S. KANEMATSU, K. ASADA. 1984. Occurrence of Cu, Zn - superoxide dismutase in the
intrathylakoid space of spinach chloroplasts.
Plant Cell Physiol. 25: 883-889.

46. HAYAKAWA, T., S. KANEMATSU, K. ASADA. 1985. Puri-
fication and characterization of thylakoid-bound
Mn - superoxide dismutase in spinach chloroplasts.
Planta 166: 111-116.

47. FOYER, C.H., B. HALLIWELL. 1976. The presence of
glutathione and glutathione reductase in chloroplasts: A proposed role in ascorbic acid metabolism.
Planta 133: 21-25.

48. KAISER, W.M. 1979. Reversible inhibition of the Calvin
cycle and activation of oxidative pentose phosphate
cycle in isolated intact chloroplasts by hydrogen
peroxide. Planta 145: 377-382.

49. LIZADA, M.C.C., S.F. YANG. 1981. Sulfite-induced lipid peroxidation. Lipids 16: 189-194.

50. TANAKA, K., K. SUGAHARA. 1980. Role of superoxide dismutase in the defense against SO_2 toxicity and induction of superoxide dismutase with SO_2 fumigation. Res. Rep. Natl. Inst. Environ. Stud. 11: 155-164.

51. RABINOWITCH, H.D., I. FRIDOVICH. 1985. Growth of Chlorella sorokiniana in the presence of sulfite elevates cell content of superoxide dismutase and imparts resistance towards paraquat. Planta 164: 524-528.

52. GRILL, D., H. ESTERBAUER, U. KLOSCH. 1979. Effect of sulphur dioxide on glutathione in leaves of plants. Environ. Pollut. 18: 187-194.

53. RABINOWITCH, H.D., D. SKLAN, P. BUDOWSKI. 1982. Photo-oxidative damage in the ripening tomato fruit: protective role of superoxide dismutase. Physiol. Plant. 54: 369-374.

54. DHINDSA, R.S., W. MATOWE. 1981. Drought tolerance in two mosses: correlated with enzymatic defense against lipid peroxidation. J. Exp. Bot. 32: 79-91.

55. CLARE, D.A., H.D. RABINOWITCH, I. FRIDOVICH. 1984. Superoxide dismutase and chilling injury in chlorella elliposidea. Arch. Biochem. Biophys. 231: 158-163.

56. CHIMENT, J.J., R. ALSCHER, P.R. HUGHES. 1986. Glutathione as an indicator of SO_2-induced stress in soybean. Environ. Exp. Bot. 26: 147-152.

57. GURI, A. 1983. Variation in glutathione and ascorbic acid content among selected cultivars of Phaseolus vulgaris prior to and after exposure to ozone. Can. J. Plant Sci. 63: 733-737.

58. BECK, E., A. BURKET, M. HOFMANN. 1983. Uptake of L-ascorbate by intact spinach chloroplasts. Plant Physiol. 73: 41-45.

59. BOYER, J.S. 1982. Plant productivity and environment. Science 218: 443-448.

60. ALSCHER, R., J. BOWER, W. ZIPFEL. 1987. The basis for different sensitivities of photosynthesis to SO_2 in two cultivars of pea. J. Exp. Bot. 38: 99-108.

Chapter Two

THE BIOCHEMISTRY OF OZONE ATTACK ON THE PLASMA MEMBRANE OF PLANT CELLS

ROBERT L. HEATH

Department of Botany and Plant Sciences
University of California
Riverside, California 92521

INTRODUCTION

Substantial research effort is currently being under-
taken towards assessing the impact of oxidant air pollutants
(generally, ozone) on economically important agricultural
crops.[1] Many of these studies rely on vast numbers of
field trials in order to determine the minimum level of
ozone which affects agricultural productivity. While
these data provide regulatory agencies with numbers to set
air quality standards, these investigations generate little
predictive information. Too many variables of the field and
crops prevent complete understanding. Air pollution studies
need predictive models but we are painfully short of them.
This paper will address the initial site of ozone injury to
cells in the hope that the process of model building can
begin.

Plants meet and react to stress in many ways. Of
course, if a portion of the plant dies because of the

29

stress, then a smaller plant, so produced, will result in
less total productivity. In many cases, however, the
stress does not kill part of the plant. Ozone-induced
stress, which will be emphasized in this review, often
leads to slight alterations in the plant's normal functioning, which can be reversed or compensated for by the
plant. In this case, homeostasis of the plant is shifted
out of balance by ozone, and energy reserves and metabolite
accumulation must be expended to restore homeostasis or to
stabilize a new level of homeostasis. This costs the plant
in terms of production of food or fiber.

Many investigators believe that diverse stresses
ultimately exhibit similar metabolic ramifications once
the cell begins to restore homeostasis (see Levitt[2] for
summary). Other reports on ozone-induced injury[3-6] have
concentrated on other metabolic aspects of this restoration.
I wish to concentrate on the first act or primary site of
ozone interaction with the cell. It is at this initial
level of disruption that the unique character of each
stress is demonstrated.

Further, this paper will not be concerned with the
entry of the pollutant into the tissue, although it is a
critical event.[3] Rather it will focus attention upon the
cells within the tissue where the ozone initially interacts
with the plant. These are the cells of the substomatal
cavity nearest to and including the guard cells and the
cells of epidermal layer. This region shows the first
visible indication of ozone damage:[4] a "water-logging"
due to a collapse of the cell's selective membrane permeability. Later these cells bleach and the surrounding tissue
turns necrotic.[3,4] Thus, this paper will focus upon the
outer cell surface — the plasma membrane — as the initial
site of injury.

Two major points should be emphasized. The first is
that ozone, while chemically reactive (see the last section
of this paper), is initially not that reactive to cells.
This has often been demonstrated by the high levels of
ozone required to cause damage to individual cells.[3,5]
This lack of damage is more clearly shown by how much
ozone must be taken up by algal cells before (1) the cells
become unable to form colonies on agar plates, and (2)
lipid oxidation products are detected within the cells.
More than 5×10^{-16} moles of ozone per nonviable cell are

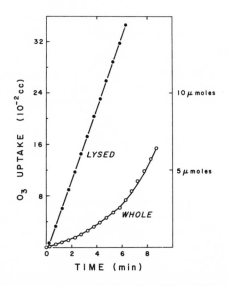

Fig. 1. Ozone uptake of red blood cells. Human red blood cells (10^8 cells/ml, total volume of 50 ml), washed in 0.18 M sucrose, 5 mM $MgCl_2$ and 10 mM tetramethyl ammonium phosphate at pH 7, were suspended either in the same medium (whole) or in 25 mM MES buffer to osmotically lyse the cells (lysed). The suspension was exposed to ozone in O_2 (180 ppm; dose, 3.9×10^{-6} moles/min ozone). The O_3 uptake was measured as previously described[8],[44] using a Cary 15 spectrophotometer which measures the difference between the entrance gas (sample) and the exit gas (reference cuvette).

removed from the gas stream before 50% of the cells fail to form colonies on plates.[8] Still another example is shown in Figure 1. Red blood cells initially do not remove much ozone from a gas stream bubbled into a solution. Considerable exposure is required before they begin to react with ozone and remove it from the gas stream, presumably by reactions with their contents, which are now exposed. On the other hand, the same cells, if lysed, remove ozone from the gas stream quite readily and nearly completely. The membrane appears to protect the cell and its contents from this reactive oxidant.

The second point is that there are at least two stages of ozone injury. The initial stage is reversible

under some conditions; necrosis does not develop although
productivity is lowered;[3,5,9] this seems to be related to a
chronic exposure, i.e. a low dose over a long time. At a
later stage, visible injury develops. This is associated
with acute (high dose over a short time) exposure. The
effects on metabolism and time courses for these two stages
are not the same and should be clearly distinguished.

ALTERATIONS IN MEMBRANE FUNCTIONS

Permeability and Potential

       Previous work from our laboratory[8,10-12] and many
others[3,5,6] has shown that membrane function is disturbed
by the introduction of ozone into the solution in which
cells are suspended. In general, the properties of the
cell related to selective permeability are changed such
that metabolites are relatively rapidly lost from the
cell.[10] Thus, ions such as potassium,[11] Rb^+ (as a tracer
for potassium),[12] and chloride[10] can no longer be maintained
within the cell and their uptake pattern is altered
depending upon the dose of the ozone. These studies were
carried out with Chlorella sorokiniana, a model system for
single cells, but similar conclusions have been reached by
others using leaf discs, after fumigation of the whole
plant.[13-15]

       The changes in the flow of $^{86}Rb^+$ under a low ozone dose
for cells exposed in water solutions is shown in Figure 2.
The lower dose slows down the normal course of events.[12]
There are three somewhat distinct phases of the change in
the movement pattern. In the first phase (noted as I),
the rate of influx for the control and treated are virtually
the same; the efflux rate is slightly higher. If we assume
that the movement of this tracer in the dark is strictly
passive and obeys the Goldman equation[16] (which bases the
ionic flow upon only the electrical potential and the
permeability coefficient), then these changes would be due
to a small depolarization (18 mV) of the membrane potential
and a low increase (10-20%) in the permeability. The
membrane potential seems to be most sensitive to the
external environment[17-19] and thus it is not surprising to
see this effect. So, there is little change with an exposure of less than 3×10^{-7} moles of ozone (for 4 minutes

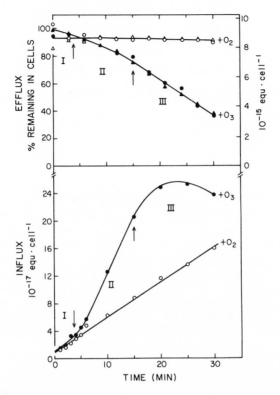

Fig. 2. The influx and efflux of potassium, as monitored by [86]Rb+ tracer, into or out of <u>Chlorella sorokiniana</u>, as influenced by ozone exposure. Cells were treated as previously described[12] except that the ozone exposure rate was considerably less (27 ppm or 7.5 x 10^{-8} moles ozone/min).

under these conditions), or an exposure of 3 x 10^{-16} moles of ozone per cell entering the system.

The second phase (II) is characterized by a 4-fold increase in efflux rate and a near doubling of the influx rate. If again we assume that the movement of this tracer obeys the Goldman equation,[16] then these changes would be due to a further depolarization (17 mV) of the membrane potential and a 3- to 4-fold increase in the permeability.

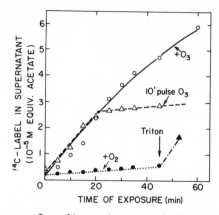

Fig. 3. The loss of radioactive metabolites from Chlorella
sorokiniana upon exposure to ozone. The experiments were
carried out as previously described,[10,12] except that the
cells were labeled with ^{14}C-acetate (5 mM at 1 µCi/ml) for
20 minutes and were washed with cold acetate solution for
10 minutes prior to the measurement of loss of radioactivity.
The ozone level was 120 ppm at a dose of 1.5×10^{-7} moles
ozone/min. Triton X-100 (0.1%) was added at 45 minutes to
the O_2-exposed cells.

    The third phase (III) is more dramatic and has been
implicated in the ozone-induced changes affecting the total
membrane structure (observed with the electron micro-
scope[20]) and in a switch from reversible to irreversible
injury.[10,12,20] Using the Goldman calculations,[16] the
membrane is now depolarized to the level normally associated
with totally damaged, yet intact cells. Here the potential
falls below -35 mV and the permeability coefficient is again
doubled to a level of nearly 10-times that of the control
cells. This third phase is caused by a dose in excess of
1.5×10^{-15} moles of ozone per cell entering.

    This near total loss in permeability properties can be
further demonstrated by using internal metabolites, labeled
by feeding ^{14}C-acetate to the cells before they are
subjected to ozone (Fig. 3). A 10-minute phase of ozone
causes an immediate loss of labeled metabolites; this loss
continues for an additional 10 minutes after the ozone
pulse is removed from the gas stream. This loss is similar

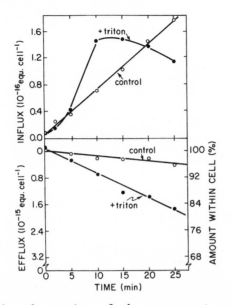

Fig. 4.   The alteration of the transport properties of
Chlorella sorokiniana upon treatment with Triton X-100.
The cells were treated as previously described,[10],[12]
with the addition of Triton (0.1%) at zero time.

to that seen after the cells, previously labeled with
acetate, are subjected to Triton X-100, a detergent and
membrane disruptant.[21]

    Triton X-100 also mimics the action of ozone quite
well using the same $^{86}Rb^{+}$ tracer studies as before[12]
(Fig. 4). Not only does Triton increase the efflux of
the internal Rb, but also induces the peculiar pattern of
influx with a short lag before the increased influx,
followed by a leveling and a later decline in accumulation.
Using the Goldman equation[16] for these data, we find that
Triton at this concentration causes a mild depolarization
(20 mV) and an increase in permeability coefficient
(approximately 3-fold). Thus, ozone acts as if it were a
detergent or, more probably, generates a detergent-like
molecule.

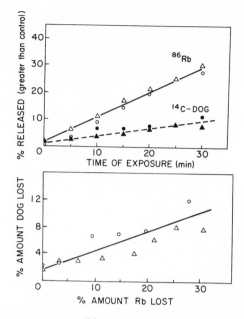

Fig. 5. The efflux of ^{86}Rb$^+$ and ^{14}C-deoxyglucose from Chlorella sorokiniana upon exposure to ozone. The algae were treated and pre-labeled as before[10],[12] with both labels. At 0 time, the cells were exposed to ozone at 109 ppm (dose of 2.73 x 10^{-7} moles of ozone/min). The percent of release is calculated on the assumption that 100% of the tracer is in the cell at t=0.

       That the membrane very rapidly becomes permeable to most, if not all, smaller molecules and ions is further demonstrated in studies of deoxyglucose transport. Deoxy-glucose is taken up by a hexose-transport system[10],[22] but seemingly is not metabolized by the cells.[13] In Figure 5 the efflux rate of deoxyglucose is less than that of ^{86}Rb$^+$, based upon the percentage of total material inside. Deoxyglucose efflux is not dependent upon membrane potential,[23] so this experiment immediately shows the extent and rapidity of the changes in permeability coefficient, without the concurrent depolarization of the membrane potential.

Before discussing the possible chemistry of the ozone interaction with the membrane, the action of ionic/pH environment upon the characteristics of the membrane should be reviewed. The cell wall surrounds the plant cell and this cell wall somewhat restricts the movement of molecules by size.[24] Furthermore, the wall itself has many characteristics, including alcoholic groups of sugars, phosphate groups and other charged groups,[25] which could affect ozone chemistry and its breakdown or reaction products. This micro-environment needs to be further understood, if we hope to understand the primary site of ozone injury.

In particular, it should be stressed that bulk pH can influence the normal transport of ions and sugars. If the pH of the solution near the wall were altered by ozone interacting with the surrounding groups, the initial events may just be due to the altered pH within the wall region and interaction with the membrane.[17-18,23] For example, both the efflux and influx of a Rb tracer can be increased by a rise in pH. When the pH of the suspending medium is increased from 6 to 8, the efflux and the influx rates increase 5- and 3-times, respectively (unpublished data). Further, the rate of deoxyglucose uptake is slowed at higher pH.[23] Thus, some of the observed early events may be due to a change in pH induced by ozone. This imbalance in pH could be readily reversed by the cell and thus would not cause the normal injury patterns. However, it would alter the "set-point" of the cell's metabolism due to the "wrong" external pH. A large change in pH could be caused by a small amount of ozone breakdown since the change occurs occurs within such a small space. For Chlorella, the wall space can be calculated to be 3.38×10^{-12} cm^3 (from electron microscopy,[20] the wall space is about 95 nm in thickness). If there is a buffer capacity of about 20 (estimation[26]), a change in pH from 6.5 to 9.5 would occur by the release of only 2.0×10^{-17} equivalents of hydroxyl ion (see later section on the breakdown of ozone in water).

It seems clear that the initial events of ozone inter-action with the cell center on the cell wall region and that reactions there would affect the normal functioning of the plasma membrane. Other interactions of ozone products or ozone itself may be outside, inside or within the membrane depending upon how far ozone or its break-down products can move before reaching biochemicals

sensitive to ozone. Certainly, disruption of normal
membrane function will drastically alter the normal
metabolism of the cell and could easily lead to gross
imbalance within the cell proper observed later in time.

Lipids and Proteins

The description of ozone attack upon the membranes
normally involves ability of ozone to break double bonds
of unsaturated fatty acids (USFA).[27] Although this concept
was discussed recently,[8] new data requires further
discussion here.

One must remember that ozone, being a very polar
molecule, is extremely hydrophilic and, under most circumstances, will not enter the region of the membrane which
contains the unsaturated fatty acids. As such, ozone
participates in reactions similar to those involving the
hydroxyl radical and solvated electron recently investigated by Barber and Thomas.[28] They showed that the
artificial bilayer, which phosphatidyl choline forms in
solution, protected the lipid from oxidative attack by
exclusion of the species from the organic phase. They
also showed that the hydroxyl radical would react with
phosphatidyl choline each time they came into contact.
To be sure, there also may be products of ozone reaction
with varied cellular biochemicals which could enter the
membrane region and react with the double bonds of USFA.
Since we know little about the interactions of these
varied products, such ideas may be speculative at best.

Under some conditions, we do see alterations of fatty
acids[29] and of polar lipid fractions.[8,30,31] I believe
that these alterations are of two types — extreme oxidations under large doses of ozone, and metabolic alterations
derived through the cell regaining homeostasis.

If the level or duration of ozone exposure is extreme
(high or long), an oxidative decomposition product —
malondialdehyde[29] — of ozone attack on USFA is observed
in most living systems investigated. More rarely in
specific systems, a loss of USFA has been observed.[29,30]
Yet, when red blood cells are exposed to ozone, no loss of
USFA could be found even after the inactivation of some
enzymes (acetyl choline esterase on the membrane surface
and glyceraldehyde-3-phosphate dehydrogenase inside the

membrane).[32] In these cases the USFA exhibited alterations
only when damage was induced by long or high exposure.

Under long term exposure at low levels of ozone,
plants show altered lipid patterns, although researchers
have not demonstrated conclusively that these lipids were
on the surface of or within the membrane.[8] Indeed several
research groups have demonstrated that the basic pattern
of lipid metabolism itself is altered.[30,31] Since the
metabolism of plants exposed to chronic levels of ozone is
quite disturbed, it is not surprising that lipid metabolism
should also be altered. But these changes in lipid seem
to be part of the general metabolic response to stress
rather than to ozone per se. Therefore, while lipid
metabolism may be altered by stress, this metabolism is
not the initial point of that stress.

In another model system, Pauls and Thompson[33] showed
that in microsomes which have been exposed to ozone (at an
average concentration of 0.2 μM in solution for 60 minutes)
the temperature at which the membranes undergo a transition
from liquid-crystalline to gel-lipid phase is increased.
Furthermore, these workers observed the formation of
distinct regions of gel-lipid phase within the membrane.
Also both the level of sterols and unsaturated fatty acids
fell in relation to the total phospholipid concentration.
This is similar to what is observed during senescence,
nicely summarized by Borochov and Faiman-Weinberg.[34] It
also fits the concept that ozone induces a premature
senescence in plants.[3,9] However, the results do not
match the increase in sterols when whole plants are fumigated with ozone as reported by Grunwald and Endress.[35]

Mudd and his coworkers have long emphasized the need
for understanding interactions of protein with ozone.[36]
Unfortunately, few investigators have concentrated their
effort on the membrane proteins, most probably due to the
difficulty in their isolation and measurement of function.
More recently, plasma membrane preparations have been made
from a large variety of plant material, although such
preparations are far from being uniform or pure.[37]

ATPase and Transport

While transport of ions in plants has been investigated for more than a century, the proteins involved have

Table 1. The ozone-induced inhibition of plasma membrane ATPases of pinto bean plants _in vivo_.

Addition		Activity (μmoles Pi/mg protein. hr)	
		Non-exposed	Ozone-treated
Mg^{++}		1.75 ± 0.39	0.0
	+DTE	1.66 ± 0.09	3.75 ± 0.12
$Mg^{++} + K^{+}$		2.92 ± 0.09	0.0
	+DTE	4.00 ± 0.07	5.86 ± 0.12

Data from Dominy & Heath.[38]

Plasma membranes were isolated by discontinuous sucrose gradient after preparation of the homogenate from either non-exposed (control) or ozone-exposed (0.5 ppm for 5 hrs) plants. Dithiothreitol (+DTE, 3 mM) was added before the assay. ATPase monitored by Pi release from ATP in the presence of Mg^{++} (3 mM) and with or without K^{+} (50 mM).

only been found and examined in the last decade. In fact, in most cases, the proteins have not been isolated in a purified form but rather as membrane fragments.[37] Yet these fragments do allow a study of the transport proteins or sites, and better understanding of their actions is at hand.

Dominy and Heath[38] were the first to use such plasma membrane preparations to investigate ozone injury to the membrane. A summary of their data is found in Table 1. Two ATPases are associated with the membrane; one is Mg^{++}-dependent and the other is K^{+}-stimulated. Under extreme conditions of ozone exposure in ozone-sensitive bean plants, both ATPases were inactivated but their activity could be restored by DTE, a thiol reagent, if the ATPases from plant material fumigated with ozone were treated after isolation. Less sensitive beans or lower exposures gave less inactivation. This reactivation by DTE suggests that the ozone acted upon sulfhydryl groups of the proteins,[37] oxidizing the sulfhydryl to a disulfide bridge.

Table 2. The ozone-induced inhibition of plasma membrane
ATPases from pinto bean in vitro.

Time of Exposure	Activity of ATPase (μmole Pi/mg protein. hr)		Level of SH* (0.1 μmole/ mg protein)
	Mg^{++}-dep.	K^+-dep.	
No exposure	6.2	3.5	6.6
20 min O_2	3.2	2.0	8.0
+DTE	6.0	3.5	–
20 min O_3	2.0	0.0	3.2
+DTE	6.6	2.8	–

*SH = sulfhydryl

Unpublished data of B. Finchk and R.L. Heath. Conditions
similar to Table 1 except that the ozone exposure (1.6 x 10^{-7}
moles of ozone/min) was carried out after the membranes were
isolated, in 5.0 ml volume (300 μg protein). Dithiothreitol
(+DTE) was added after 20 min of exposure and the rate was
remeasured 10 min later. S.E. for all data was 0.6 to 0.7.

The dose of ozone required to inactivate the ATPase
can be estimated.[39] Under well-watered conditions,
stomatal conduction of pinto bean is about 0.7 cm/sec;
thus, the rate of ozone entry into the leaf under the
above fumigation conditions[38] (at 0.5 ppm for 5 hours)
would be 2.6 x 10^{-7} moles of ozone absorbed per cm^2 leaf
area. For a ratio of 12 for the internal leaf area to
external leaf area,[40] the total ozone dose would be about
2.4 x 10^{-8} moles ozone/cm^2 cell surface. This is similar
to what is observed for the Chlorella system and movement
of Rb^+ ions (calculated from the above data for dose per
cell and the surface area of Chlorella of 3.56 x 10^{-7}
cm^2/cell).

This work was confirmed in vitro by B. Finchk and
R.L. Heath (unpublished data). Table 2 shows that both
ATPases from isolated plasma membrane fragments are
inactivated by ozone exposure (at a dose of 3.2 x 10^{-6} moles
ozone or 1.1 x 10^{-15} moles/mg protein) but not to the same

extent; the K^+-dependent ATPase is more sensitive. The
activity of both can be partially restored by a thiol
reagent and upon exposure to ozone there is a drop in
total measured sulfhydryl groups within the preparation.

Thus, it is clear that the ATPases which are thought
to be associated with ion pumps on the membrane can be
inactivated by ozone and this inactivation may be due to
a simple oxidation of the sulfhydryl to a disulfide bridge.
Also it should be again noted that the Hexose transport
system has a sulfhydryl necessary for activity.[22]

CHEMICAL REACTIVITY IN WATER

The literature is confusing regarding the types of
reactions that ozone can undergo in aqueous solutions.
Reactions of ozone in organic solvents are well understood
and useful for organic synthesis;[41] however, when the
solvent can participate in the chemistry of ozone, the
picture is not so clear. In fact, confusion has arisen
in the past because some researchers have not paid
sufficient attention to the diversity of reactions that
are possible.

In the leaf ozone enters as a gas but rapidly reaches
a region where the relative humidity is 100%. The ozone
molecule must enter solution as it reaches the individual
cell and penetrates into the wall region. This area is
poorly defined chemically at the present time, but certainly
is rich in materials which can dramatically influence the
paths by which ozone breaks down. Thus, it is critical to
understand ozone's reactivity in order to predict the
initial site of ozone attack. In this section, three
biochemical systems likely to react with ozone are
described: (1) the aqueous solution itself; (2) unsaturated fatty acids; and (3) sulfhydryl groups. There are
other biochemical species present, but the ones cited
are currently believed to be the chemicals associated
with the active site.

Water Solution

The reactions of ozone in water have had a confusing
history ever since Weiss[42] first described their reactions
and postulated a mechanism. Since that time, the reports

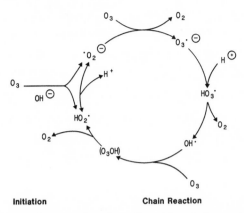

Initiation   **Chain Reaction**

Fig. 6. The reaction mechanism for the interaction of
ozone within a water solution, as modified from Staehelin
and Hoigne.[45,46] Hydroxyl radical (HO·), peroxyl radical
(HO$_2$·), and superoxide radical (O$_2^-$·) as indicated.

have been numerous and it is doubtful that the final
picture has emerged. Yet, some points of agreement are
presented in Figure 6. Ozone in acid solution is reason-
ably stable; its breakdown can hardly be measured.[43-45]
As the solution becomes more alkaline, a hydroxide ion
reaction becomes significant and leads into a "cyclic"-
type of reaction,[45,46] in which the first products are the
protonated and unprotonated peroxy-radicals HO$_2$ and O$_2^-$,
the latter superoxide anion formed by dissociation of HO$_2$
(pK=4.8). The initiation is highly pH-dependent and
generates radicals, which probably has caused the diverse
radical reactions, often noted in the literature.[41,44,47]
The peroxyl-radicals have been long noted as inducers or
initiators of peroxidation[47] of unsaturated fatty acids
(USFA).

Several apparent contradictions occur with regards to
the entry of ozone into solution. The solubility of ozone
is often listed as greater in acid solution.[46,48] In a
sense it is; however, alkaline solution catalyzes the
breakdown of ozone and so lowers the measured steady-state
concentration of ozone.[45,49] Indeed, when the change in
pH induced by ozone is examined,[48] one observes a greater
pH change (release of hydroxyl ion or uptake of protons)

Table 3. Concentration of various oxidizing species in water derived from the interaction of water and ozone in solution.

Species	Concentration (M)	
	pH7	pH9
Superoxide radical ($O_2 \cdot$)	8.75×10^{-15}	1×10^{-12}
Ozone radical ($O_3 \cdot$)	4.16×10^{-15}	5×10^{-14}
Protonated ozone radical ($HO_3 \cdot$)	1.48×10^{-16}	1×10^{-18}
Rate of ozone loss	0.015%/min	0.21%/min
Yield (hydroxyl radical)*	11%	0.008%

*Amount of radical produced/ozone consumed in reaction.

Calculated from the data of Staehelin and Hoigne[46] on the basis of 100 ppm ozone in a gas stream.

at lower pH. This suggests that the protonation of ozone upon solubilization releases a hydroxyl ion.[48,49] This is the rate-limiting step only at acidic pH; the pH decline with added ozone lessens at more alkaline pH.[48]

The reactions of the cycle are less well understood, but reasonable with respect to products and kinetics.[45,46] A second ozone reaction with the superoxide radical generates oxygen and an ozone radical, which reacts with protons to break down rapidly to hydroxyl radical, a very reactive species which has been detected by spin-trapping under certain circumstances.[47,50] The cycle is postulated to continue with a third ozone reaction by regenerating peroxyl radical. There are many places where a variety of chemicals can either promote or inhibit (by radical scavenging) the reactions.[46] Promoters include carboxylic acids, alcohols (e.g., sugars), and bicarbonate/CO_2. The biologically ubiquitous ion, phosphate, has been reported to be a possible scavenger.[45]

If we use the kinetic parameters from Staehelin and Hoigne,[45,46] we can calculate the values of the concentration of various species (Table 3). We assume that steady-state is reached rapidly and that hydroxyl radical reacts

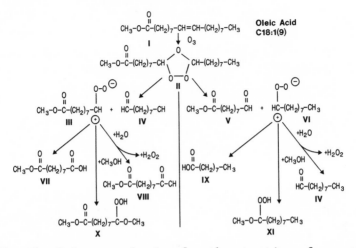

Fig. 7. The Criegee mechanism for the reaction of ozone with the unsaturated fatty acid, oleic acid (C18:1Δ^9), adapted.[27,41,51]

readily with other species and does not build up to complete the cycle. At pH 7, the levels of these species are very low and the loss of ozone in solution is only 0.015% of that in the solution, similar to what is observed.[46] The yield of hydroxyl radical production (compared to the amount of ozone breakdown) is relatively high, 11%. As the pH rises to 9, the breakdown of ozone in solution increases (to 0.21%) while the yield of hydroxyl radical drops to less than 0.01%. This complex reaction sequence with varied effectors may explain why the kinetics of ozone breakdown has been so controversial in the past.[3,7,45]

Unsaturated Fatty Acids

The history of ozone attack on double bonds has also been filled with controversy.[3,7,40,47] The accepted mechanism for ozone/double-bond interaction is known as the Criegee reaction[27,41] in which the double bond is broken and the ozone molecule is inserted as a peroxyl and ether bridge (see Bailey[41] for the mechanism) (Fig. 7). This species (II) is unstable and rapidly rearranges, especially in polar solvents. Species III and VI appear to form

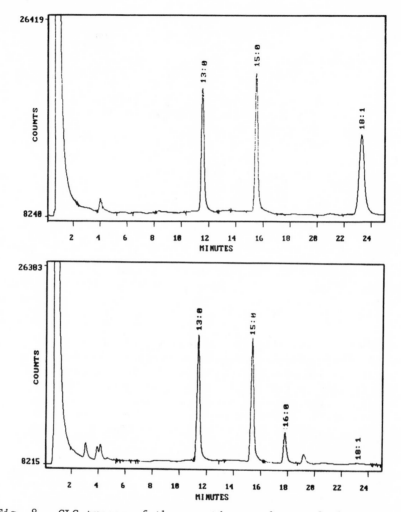

Fig. 8. GLC traces of the reaction products of oleic acid with oxygen and ozone. Methyl ester of oleic acid (2.75 µg/50 ml) was suspended in methanol and subjected to 15 minutes of pure oxygen (top curve) or 180 ppm ozone in oxygen (bottom curve, 1.12×10^{-6} moles ozone/min). The samples were not extracted in water but concentrated and run on a 15% DEGS column as previously described.[29] Unknowns are tabulated according to retention time (shown in brackets, minutes) and are U1 (2.75), U2 (4.69), U3 (17.64), and U4 (19.00). U3 has a retention time similar to C16:0, but seems not to be it. The internal standards are C13:0 and C15:0.

H_2O_2 in water[41,47,51] and an organic hydroperoxide in alcohol.[47,51]

We have attempted to isolate these products (especially the aldehydes and carboxylic acid) using standard GLC. We were successful if we used an alcohol as the solvent, since we did not have to reextract the solution with a water fraction for the fragments (Fig. 8). Unfortunately, if the gassing were carried out in phosphate buffer (with a final extraction), we lost the species, presumably because they were water-soluble. Short chain fatty acids are water-soluble[52] and aldehydes are even more so. There are, however, three new peaks which are unique to normal USFA found in plants (Fig. 8, peaks U1, U2, U3). We have not yet fully identified them but their positions are consistent with previous work on USFA oxidation products. We hope to be able to extract lipids with one portion of the short chain length aldehyde from plants or plasma vesicles exposed to ozone in order to confirm these reactions in vitro.

The reaction scheme is highly sensitive to water, as shown by Heath and Tappel.[51] Table 4 is a summary of their data which shows a high rate of production in H_2O_2 in aqueous solution, indicative of the participation of the solvent in the reactions (Fig. 7, species IV and VIII). Yet if ethanol is present, a hydroperoxide is produced (which does not react with catalase) and only 1/3 of the H_2O_2 is produced. The solvent only causes a differential reaction of ozone with the fatty acid — ethanol increases the total amount of peroxide formed.

Sulfhydryl Reactions

As we have seen, ozone apparently alters a critical sulfhydryl group in the ATPase of the plasma membrane and inactivates the enzyme. Yet this is an oxidation which does not seem to proceed beyond the disulfide state,[36] since it can be reduced by the thiol reagents[38] (also see Table 2). This has not been previously observed. It was believed that sulfhydryl groups are oxidized to a sulfone,[41] which can not be reduced by thiol reagents. On the other hand, earlier work by Heath et al.[8] suggested that the kinetics of ozone uptake by sulfhydryls did not indicate such a complex oxidation.

Table 4.   The production of hydrogen peroxide and organic
peroxides from reactions of linoleic acid with ozone.

Conditions	Linear Rate $(10^{-7}$ moles/min)	
	EtOH	Water
Ozone introduced	3.07	3.01
Total peroxide	2.11	1.22
Hydrogen peroxide	0.69	1.22
Ratio (ROOH/O$_3$)*	69%	41%
Ratio (HOOH/ROOH)	33%	100%

*ROOH = total peroxide, organic and hydrogen peroxide.

Data from Heath and Tappel.[51]

Linoleic acid (150 μM) was gased with 35 ppm ozone in 95%
ethanol solution (EtOH), or phosphate buffer, pH 6.8,
(water) for 10 minutes.   The linear rate is that rate
after the initial lag.

        If we re-examine the types of reactions that can occur
in solution, this quandary can be understood.   The various
initial radicals produced by hydroxide ion[45,46] will
rapidly react with reduced compounds, such as sulfhydryls,
to neutralize the radical and convert the superoxide
radical into a superoxide molecule.   This produces a
thiol radical which can react with another one to produce
a disulfide bridge, and yield second order kinetics with
respect to sulfhydryl concentration.[8]   Thus, sulfhydryl
reactions should occur at higher pH and be sensitive to
reagents which can react with peroxyl radicals more
rapidly than the sulfhydryls.

CONCLUSION

        At this stage it is impossible to describe accurately
what initial reactions of ozone near the membrane produce

the changes observed in its semipermeable characteristics.
I believe that the ozone molecule in its reactions with
water, modified by various groups within the wall, changes
the microenvironment near the membrane. The "sensors" of
the environment within the membrane give the regulatory
agents within the cell the wrong signal. The cell's
metabolism responds to incorrect signals and the cell's
homeostasis is distorted. The cell is reacting to a
situation which does not exist in the whole plant. This
response is reversible and one that corrects itself once
the flow of ozone ceases. This response is also subtle
and will be difficult to measure.

This is not the only reaction which occurs however.
External groups such as sulfhydryls may be directly
modified by ozone causing a decline in transport. As
the semipermeable nature and transport capacities of the
membrane change, perhaps by a fluidity change,[34] ozone
can penetrate deeper into the membrane itself where it
or one of its reaction products can react with membrane
components. Given their reactive nature, buried sulfhydryl
groups and unsaturated fatty acids become the targets.
This leads to a further breakdown of membrane organization, leading ultimately to membrane vesiculation[20] and
finally lysis and a release of the cell's contents (e.g.,
Fig. 1).

The cell's response to these oxidations is to mobilize
the antioxidants within the cell in order to reduce the
oxidized species and to break all radical-induced chain
reactions. But these responses are another story that
hopefully will lead to an understanding of how reversible
ozone-induced injury is and how successful the cell will
be in reversing these later, more damaging events.

It is clear that a more complete understanding of the
reactions of ozone in water solutions is needed, in order
to predict successfully what biochemicals are at risk in
ozone injury. In addition to being able to predict what
sites are most sensitive, we may be able to understand what
events in metabolic pathways are critical in the protection
of the cell and plant.

REFERENCES

1. HECK, W.W., O.C. TAYLOR, R. ADAMS, G. BINGHAM, J.
   MILLER, E. PRESTON, L. WEINSTEIN. 1982.
   Assessment of crop loss from ozone. J. Air
   Pollut. Control Assoc. 32: 353-361.
2. LEVITT, J. 1972. Responses of Plants to Environmental Stress. Academic Press, New York, 697 pp.
3. HEATH, R.L. 1980. Initial events in injury to
   plants by air pollutants. Annu. Rev. Plant
   Physiol. 31: 395-431.
4. GLATER, R.B., R.A. SOLBERG, F.M. SCOTT. 1962. A
   developmental study of the leaves of Nicotiana
   glutinosa as related to their smog-sensitivity.
   Am. J. Bot. 49: 954-970.
5. ADAMS, R., J.H.B. GARNER, L.W. KRESS, J.A. LAURENCE.
   R. OSHIMA, E. PELL, R. REINERT, O.C. TAYLOR,
   D.T. TINGEY. 1986. Effects of ozone and other
   photochemical oxidants on vegetation. Air Quality
   Criteria for Ozone and Other Photochemical
   Oxidants. EPA Document EPA-600/8-84-020A,
   Chapter 7.
6. KOZIOL, M.J., F.R. WHATLEY, eds. 1984. Gaseous Air
   Pollutants and Plant Metabolism, Butterworth,
   London, 466 pp.
7. HEATH, R.L. 1984. Air pollution effects on
   biochemicals derived from metabolism: organic,
   fatty and amino acids, ibid., pp. 275-290.
8. HEATH, R.L., P. CHIMIKLIS, P. FREDERICK. 1974.
   The role of potassium and lipids in ozone
   injury to plant membranes. In Air Pollution
   Effects on Plant Growth. (W.M. Dugger, ed.),
   Am. Chem. Soc. Series No. 3, pp. 58-75.
9. McCUNE, D.C., L.H. WEINSTEIN, D.C. MacLEAN, J.S.
   JACOBSON. 1967. The concept of hidden injury
   in plants. In Agriculture and the Quality of
   Our Environment. (C. Brady, ed.), Am. Assoc.
   Adv. Sci. USA, Washington, D.C., pp. 33-44.
10. HEATH, R.L. 1984. Decline in energy reserves of
    Chlorella sorokiniana upon exposure to ozone.
    Plant Physiol. 76: 700-704.
11. CHIMILKLIS, P.E., R.L. HEATH. 1975. Ozone-induced
    loss of intracellular potassium ion from
    Chlorella sorokiniana. Plant Physiol. 56: 723-
    727.

12. HEATH, R.L., P.E. FREDERICK. 1979. Ozone alteration of membrane permeability in Chlorella I: permeability of potassium ion as measured by [86]Rubidium tracer. Plant Physiol. 64: 455-459.

13. PERCHOROWICZ, J.T., I.P. TING. 1974. Ozone effects on plant cell permeability. Am. J. Bot. 61: 787-793.

14. SUTTON, R.M., I.P. TING. 1977. Evidence for repair of ozone induced membrane injury. Atmos. Environ. 11: 273-275.

15. EVANS, L.S., I.P. TING. 1974. Effect of ozone on ^{86}Rb-labeled potassium transport in leaves of Phaseolus vulgaris. Atmos. Environ. 8: 855-861.

16. GOLDMAN, D.E. 1943. Potential, impedance and rectification in membranes. J. Gen. Physiol. 27: 37-60.

17. LEFEBVRE, J., C. GILLET. 1973. Effect of pH on the membrane potential and electrical resistance of Nitella flexilis in the presence of calcium. J. Exp. Bot. 24: 1024-1030.

18. TROMBALLA, H.W. 1974. Der einfluss des pH werts auf aufnahme und abgabe von natrium durch Chlorella. Planta 117: 339-348.

19. ZIMMERMANN, U., F. BEAKERS. 1978. Generation of action potentials in Chara corallina by Turgor pressure changes. Planta 138: 173-179.

20. SWANSON, E.S., M. TOIVIO-KINNUCAN, R. HEATH, W.P. CUNNINGHAM. 1982. Ozone-induced ultrastructural changes in the plasma membrane of Chlorella sorokiniana. Plant, Cell Environ. 5: 375-383.

21. HEATH, R.L. 1975. Conversion of signals from ion-specific electrodes to linear concentrations. Plant Physiol. 56: 181-184.

22. KOMOR, E., H. WEBER, W. TANNER. 1978. Essential sulfhydryl group in the transport-catalyzing protein of the hexose-protein cotransport system. Plant Physiol. 61: 785-786.

23. SCHWAB, W.G.W., E. KOMOR. 1978. A possible mechanistic role of the membrane potential in proton-sugar cotransport of Chlorella. FEBS Lett. 87: 157-160.

24. CARPITA, N., D. SABULARSE, D. MONTEZINOS, D.P. DELMER. 1979. Determination of the pore size of plant walls of living plant cells. Science 205: 1144-1147.

25. DUGGER, W.M., S. BARTNICKI-GARCIA, eds. 1984.
    Structure, Function, and Biosynthesis of Plant
    Cell Walls. Am. Soc. Plant Physiol./UCR Sympos.,
    505 pp.
26. NIEBOER, E., J.D. MacFARLANE. 1984. Modification of
    plant cell buffering capacities by gaseous air
    pollutants. In M.J. Koziol, F.R. Whatley, eds.,
    op. cit. Reference 6, pp. 313-329.
27. CRIEGEE, R. 1975. Mechanism of ozonolysis. Angew.
    Chem. (intl. ed.) 14: 745-760.
28. BARBER, D.J.W., J.R. THOMAS. 1978. Reactions of
    radicals with lecithin bilayers. Radiat. Res. 74:
    51-65.
29. FREDERICK, P.E., R.L. HEATH. 1975. Ozone-induced
    fatty acid and viability changes in Chlorella.
    Plant Physiol. 55: 15-19.
30. SWANSON, E.S., W.W. THOMSON, J.B. MUDD. 1973. The
    effect of ozone on leaf cell membranes. Can. J.
    Bot. 51: 1213-1219.
31. FONG, F., R.L. HEATH. 1981. Lipid content in the
    primary leaf of bean (Phaseolus vulgaris) after
    ozone fumigation. Z. Pflanzenphysiol. 104: 109-
    115.
32. FREEMAN, B. 1978. The Effects of Ozone on Human
    Erythrocytes and Phospholipid Vesicles. Ph.D.
    Thesis, Univ. Calif. Riverside, 111 pp.
33. PAULS, K.P., J.E. THOMPSON. 1981. Effects of in
    vitro treatment with ozone on the physical and
    chemical properties of membranes. Physiol.
    Plant 53: 255-262.
34. BOROCHOV, A., R. FAIMAN-WEINBERG. 1984. Biochemical
    and biophysical changes in plant protoplasmic
    membranes during senescence. What's New Plant
    Physiol. 15: 1-4.
35. GRUNWALD, C., A.G. ENDRESS. 1985. Foliar sterols
    in soybean exposed to chronic levels of ozone.
    Plant Physiol. 77: 245-247.
36. MUDD, J.B., S.K. BANERJEE, M.M. DOOLEY, K.L. KNIGHT.
    1984. Pollutants and plant cells: effects on
    membranes. In M.J. Koziol, F.R. Whatley, eds.,
    op. cit. Reference 6, pp. 105-116.
37. SERRANO, R. 1985. Plasma Membrane ATPase of Plants
    and Fungi. CRC Press, Inc., Boca Raton, Florida,
    174 pp.

38. DOMINY, P.J., R.L. HEATH. 1985. Inhibition of the K$^+$-stimulated ATPase of the plasmalemma of pinto bean leaves by ozone. Plant Physiol. 77: 43-45.

39. HILL, C.A. 1971. Vegetation: a sink for atmospheric pollutants. Air Pollut. Control Assoc. J. 21: 341-346.

40. TICHA, I., J. CATSKY. 1977. Ontogenetic changes in the internal limitations to bean-leaf photosynthesis. Photosynthetica 11: 361-366.

41. BAILEY, P.S. 1978. Ozonation in Organic Chemistry: Olefinic Compounds, Vol. 1, Academic Press, New York, 493 pp.

42. WEISS, J. 1935. Investigations on the radical HO$_2$ in solution. Trans. Faraday Soc. 31: 668-681.

43. SHECHTER, H. 1973. Spectrophotometric method for determination of ozone in aqueous solutions. Water Res. 7: 729-739.

44. HEATH, R.L. 1979. Breakdown of ozone and formation of hydrogen peroxide in aqueous solutions of amine buffers exposed to ozone. Toxicol. Lett. 4: 449-453.

45. STAEHELIN, J., J. HOIGNE. 1982. Decomposition of ozone in water. Rate of initiation by hydroxide ion and hydrogen peroxide. Environ. Sci. Technol. 16: 676-681.

46. STAEHELIN, J., J. HOIGNE. 1985. Decomposition of ozone in water in the presence of organic solutes acting as promoters and inhibitors of radical chain reactions. Environ. Sci. Technol. 19: 1206-1213.

47. PRYOR, W.A., J.W. LIGHTSEY, D.G. PRIER. 1982. Production of free radicals in vivo for the oxidation of xenobiotics: the initiation of autooxidation of polyunsaturated fatty acids by NO$_2$ and O$_3$. In Lipid Peroxides in Biology and Medicine. (K. Yagi, ed.), Academic Press, New York, pp. 1-22.

48. SELM, R.P. 1959. Ozone oxidation of aqueous cyanide waste solution in stirred batch reactors and packed towers. Adv. Chem. Ser. 21: 66-77.

49. GUROL, M.D., P.C. SINGER. 1982. Kinetics of ozone decomposition: a dynamic approach. Environ. Sci. Technol. 16: 377-383.

50. GRIMES, H.D., K.K. PERKINS, W.F. BOSS. 1983. Ozone degrades to hydroxyl radical under physiological

conditions: a spin trapping study. Plant
Physiol. 72: 1016-1020.

51. HEATH, R.L., TAPPEL, A.L. 1976. A new sensitive
assay for the measurement of hydroperoxides.
Anal. Biochem. 76: 184-191.

52. TAYLOR, M.A., L.H. PRINCEN. 1979. Fatty acids in
solution. In Fatty Acids. (E.H. Pryde, ed.),
Am. Oil Chem. Soc., Champaigne, Illinois, pp.
195-211.

53. AOSHIMA, H. 1973. Analysis of unsaturated fatty
acids and their hydroperoxy- and hydroxyl-
derivatives by high performance liquid chromatography. Anal. Biochem. 87: 49-55.

Chapter Three

EFFECTS OF OZONE AND SULFUR DIOXIDE STRESS ON GROWTH AND
CARBON ALLOCATION IN PLANTS

JOSEPH E. MILLER

USDA/ARS-NCSU Air Quality - Plant Growth
and Development Research Program
1509 Varsity Drive
Raleigh, North Carolina 27606

INTRODUCTION

Of the gaseous air pollutants that impact the United
States, ozone (O_3) and sulfur dioxide (SO_2) probably have
the greatest potential for damage to vegetation. A large
base of information on the phytotoxicity of these gases
has accumulated over the last 30-40 years. Since a
thorough discussion of all aspects of this research is not
possible, this review will emphasize certain aspects of
the effects of O_3 and SO_2 on growth and carbon allocation
(e.g., photosynthesis, partitioning and metabolism).

Even within this area, there is much information, so that a
certain amount of selectivity was necessary and a number of
excellent works were not included. A number of books and
reviews are available on the topic of toxicity of O_3 and
SO_2 to vegetation.[1-11]

    Ozone is a secondary pollutant that arises from
photochemical reactions in the atmosphere. Nitrogen oxides
(NO_x) and hydrocarbons are considered to be the major
primary emission products contributing to the buildup of
anthropogenic O_3. Some "natural" O_3 occurs in the tropo-
sphere due to stratospheric intrusion, and in some cases
natural biogenic hydrocarbons may contribute to O_3
formation. The exact extent to which these natural sources
contribute to current tropospheric levels of O_3 is not
certain, but most estimates put it at less than 50% under
O_3 polluted conditions. The major sources of NO_x are
transportation and stationary fuel combustion, each
accounting for nearly one-half of the total emissions.[12]
Emission trends for NO_x increased until about 1972 and have
been relatively stable since then. A slight increase is
expected through the year 2000. Transportation is also a
major source of hydrocarbons, although a variety of
industrial processes and other forms of combustion
contribute significantly. Emission trends for hydrocarbons
also have been fairly stable since the 1970s and no major
changes are expected. Most NO_x is emitted as NO and then
oxidized to NO_2; in the presence of sunlight, hydrocarbons
and O_2, NO_2 forms O_3. Although most sources of NO_x and
hydrocarbons are near urban areas, the formation of O_3 may
occur hundreds of miles from major emission sources due to
atmospheric transport of the precursors. Thus, a large
part of the eastern U.S. and certain areas in the midwest,
south and west are heavily impacted with O_3 when the proper
conditions occur. Because O_3 formation requires sunlight,
the general diurnal trend is for O_3 concentrations to
climb in the morning, peak for 4-6 h in the afternoon and
decrease at night.

    The major source of SO_2 in the U.S. is stationary fuel
combustion; it accounted for 82% of the emissions in 1978.
In that year other industrial processes accounted for 15%
and transportation for the remaining 3%. The total
emissions in 1978 were an estimated 27 million tons.[13]
Ninety-five percent of these emissions occurred in the
eastern half of the U.S. Recent projections suggest that

SO_2 emissions will not change substantially in the U.S. or
Canada through the year 2000.[14] It is estimated that under
average conditions about 2% h^{-1} of SO_2 is oxidized to
sulfate, although this can vary widely depending on the
other components of the air such as O_3 and metals. With
the use of tall stacks on utilities and other emission
sources to reduce local impacts, there has been a tendency
for longer-range transport in the atmosphere. However, the
major impacts of gaseous SO_2 on plants in the U.S. are
probably still located fairly near the source of the
pollution, because dilution and conversion during transport
generally reduce SO_2 concentrations to non-phytotoxic
levels.

GROWTH, BIOMASS AND YIELD EFFECTS

    In this section, research on the effects of O_3 and
SO_2 on growth and yield are highlighted. Considerable work
has been done in this area, so the discussion is necessarily
selective. Visible injury as a result of O_3 or SO_2 stress
will not be emphasized because its correlation with growth
and yield is often poor.[15] Experiments in controlled
environments and in greenhouses will be mentioned, but the
main emphasis will be on field studies because these provide
a more realistic view of the actual consequences of O_3 and
SO_2 pollution. This does not imply that controlled
environment or greenhouse studies are less important; they
are essential to the study of mechanisms of pollution
action and to the understanding of interactions of environmental factors with air pollution stress.

Ozone Effects

    Ozone affects the growth of most plants that have been
studied, and several excellent reviews have been published
on this subject.[3,9,11,15,17] Most of the species studied
to date have been annual agronomic or horticultural species.
This is partly due to their economic importance, but also
because of their convenience for experimental work (e.g.,
fast growing, genetically uniform and easy to maintain).
There are many experimental difficulties in working with
perennials such as trees. Considerable research has been
done with seedlings but also in some cases with mature
trees. It is clear from the research done to date that

many species are affected by O_3 concentrations that often occur in the U.S.

Some of the most extensive field experiments on the effects of air pollution on plants have been conducted by the National Crop Loss Assessment Network (NCLAN). This program has been funded since 1980 by the United States Environmental Protection Agency for the purpose of providing research data for assessing air pollutant (primarily O_3) damage to crops in the U.S. The general approach of this program has been to support field research in different regions of the country to obtain dose-response data on crops of the greatest economic importance on a regional and national basis. A standard protocol, using open-top chambers in the field, is used at all sites. These chambers allow investigators to control the crop exposure to O_3 throughout the growing season under conditions that are similar to normal field conditions. For an in-depth discussion of open-top chambers and the NCLAN program see Heagle et al.[18] and Heck et al.[19] Data generated by the NCLAN sites are used in developing response models for assessments of current and potential impacts of air pollution on crop production.[20]

Approximately 15 crop species have been studied in the NCLAN program and the yields of many are reduced by ambient levels of O_3 in regions where they are grown (Table 1). As explained earlier, O_3 is a regional pollutant that exerts a chronic stress throughout the growing season. For example, in the midwestern and eastern U.S., the mean O_3 concentration for 7 h day^{-1} (0900-1600 ST) during the growing season typically ranges from 0.04 μL L^{-1} to 0.06 μL L^{-1} with periods when concentrations are often as high as 0.10 to 0.15 μL L^{-1} on a daily basis. Thus, studies such as those shown in Table 1 indicate that ambient O_3 may reduce yields of crops such as cotton (Gossypium hirsutum), soybean (Glycine max), peanut (Arachis hypogea) and wheat (Triticum aestivum) by as much as 10 to 30% in some regions. It seems that the degree of response to O_3 may also vary from year to year (e.g., compare the 'Davis' soybean data and the Acala SJ-2 cotton data in 1981 and 1982). Environmental factors can interact significantly with O_3 stress; soil water status and evaporative demand have been suggested as major factors since they affect stomatal function and therefore control gas exchange between the leaf and atmosphere. It is probable that there are significant

Table 1. Predicted percent yield losses at four seasonal levels of oxide (O_3).[a]

| Species, 'cultivars' | O_3 Concentration ($\mu L\ L^{-1}$)[b] | | | |
	0.04	0.05	0.06	0.09
Barley				
'Poco'[138]	0.1	0.2	0.5	2.9
Bean, kidney				
'Calif. light red'[139]	11.0	18.1	24.8	42.6
Corn[140]				
'PAG 397'	0.2	0.7	1.5	8.1
'Pioneer 3780'	1.2	2.6	4.8	16.7
Cotton				
'Acala SJ-2' (81)[141]	5.9	10.0	14.0	25.9
'Acala SJ-2' (82)[141]	11.3	20.9	31.4	62.4
'Stoneville 213'[c]	4.8	9.9	16.4	42.2
'McNair 235'[d]	8.1	17.0	30.1	78.2
Peanut				
'NC-6'[142]	6.4	12.3	19.4	44.5
Sorghum				
'De Kalb 28'[143]	0.8	1.5	2.5	6.5
Soybean				
'Amsoy'[d]	5.5	9.4	14.3	34.3
'Corsoy' (83)[144]	5.6	10.4	15.9	34.8
'Corsoy' (86)[d]	3.5	6.8	11.7	35.2
'Davis' (81)[35]	11.5	18.1	24.1	39.0
'Davis' (82)[145]	5.1	9.8	15.4	35.6
'Forest'[146]	1.3	2.8	5.0	15.3
'Hodgson'[147]	10.3	16.6	22.4	--
'Williams'[146]	5.6	9.9	14.5	29.1
Winter wheat[148]				
'Abe'	3.1	6.2	10.2	26.7
'Arthur'	3.7	7.2	11.4	27.4
'Roland'	9.4	16.4	23.7	45.4

[a] Yield losses predicted using the Weibull Function, Yield = α exp $[-(x/\sigma)^c]$ where x = O_3 concentration, α = the hypothetical maximum yield at zero O_3 concentration, σ = the O_3 concentration when yield is 0.37 α and c = the shape parameter. Cited authors are original source of data; calculations are from Heck et al.[149]

[b] Seasonal 7-h mean (0900-1600 ST) during growing seasons.

[c] In press

[d] Personal communication, Lance Kress, Argonne National Laboratory, Argonne, Illinois (Corsoy); Joseph E. Miller, USDA, Raleigh, North Carolina (McNair 235).

cultivar or variety differences in O_3 sensitivity within a species, although this has not been thoroughly tested under the same conditions in these experiments. Other evidence does strongly suggest that cultivar differences may be very important.[21]

In the previous discussions, the effects of O_3 on yield were emphasized. In order to understand more about the mechanisms of O_3 stress on plants, studies have been done to determine the effects of O_3 stress on partitioning of biomass among tissues; one approach has been the application of growth analysis techniques. For example, Endress and Grundwald[22] found that chronic O_3 stress at concentrations similar to those in O_3-impacted areas reduced several indices of plant growth. Among those parameters reduced were: total plant dry weight, leaf area, leaf area duration, relative growth rate of leaves, stems and roots, and root/shoot ratios. Overall, these results indicate that O_3 altered the production and distribution of photosynthates within the plant. A number of other studies using a variety of experimental approaches have often led to similar conclusions (Table 2). One significant generality evident from these studies is that O_3 stress often suppresses root growth proportionally more than shoot growth. This has been attributed to the metabolic costs associated with repair processes in leaves at the expense of translocation to roots and other tissues. Isotope studies with photoassimilated ^{14}C- and ^{13}C-labelled CO_2 also have suggested a greater retention of labeled C in shoots at the expense of translocation to roots. Blum et al.[23] found that after six days of exposure of white clover (Trifolium repens) to O_3 concentrations up to 0.10 $\mu L \ L^{-1}$, ^{14}C allocation to roots increased but at 0.15 $\mu L \ L^{-1}$ of O_3 the allocation to roots was sharply decreased with a concomitant increase in allocation to developing leaves. Okano et al.[24] used $^{13}CO_2$ labelled bean (Phaseolus vulgaris) and found a reduction in ^{13}C transport to roots and an increase in transport to young leaves. McLaughlin and McConathy[25] found a similar retention of ^{14}C in leaves of O_3-exposed bean while developing pods contained less label than controls. Studies of carbohydrate levels in shoot and root tissue have led some investigators to similar conclusions. For example, Tingey et al.[26] found that O_3 increased soluble sugar and starch levels in the tops of ponderosa pine (Pinus ponderosa) and decreased levels in the roots. The relationship of pool levels of metabolites to plant development is often difficult to interpret since it represents a balance between various anabolic and catabolic reactions, as well as translocation within the plant. However, isotope studies and metabolite analyses both help describe O_3-induced changes in carbon-partitioning within plants.

Table 2.   Effect of Ozone (O_3) on partitioning of biomass.

Species	O_3 ($\mu L\ L^{-1}$)	Exposure duration	Plant parts (Effect)[1]
Soybean (Glycine max)[150]	0.10	8 h day^{-1}; 15 days	S(D); R(D); R/S(D)
Sunflower (Helianthus annuus)[151]	0.1 and 0.2	continuous; 12 days	L(D); S(D); R(D); R/S(D)
Potato (Solanum tuberosum)[152]	$\leq$ 0.27	continuous; 120 days	L(D); S(D); O(D);[a] R(D); R/S(D)
Carrot (Daucus carota)[153]	0.19 & 0.25	6 h day^{-1}; 23 days	L(N); O(D); R(D)
Fescue (Festuca arundinacea)[38]	0.1	6 h day^{-1}; 12 days	S(N); O(D);[b] R(D); R/S(D)
	0.2		S(D); O(D);[b] R(D); R/S(D)
	0.3		S(D); O(D);[b] R(D); R/S(D)
Bean (Phaseolus vulgaris)[154]	0.1	continuous; 7 days	O(D);[c] R/S(N)
	0.2		O(D);[c] R/S(D)
	0.4		O(D);[c] R/S(D)
Clover (Trifolium incarnatum)[155]	0.03	8 h day^{-1}; 42 days	L(N); O(N);[c] R(D); R/S(D)
	0.09		L(D); O(D);[c] R(D); R/S(D)
Ryegrass (Lolium multiflorum)[155]	0.03	8 h day^{-1}; 42 days	L(N); O(N); R(D); R/S(N)
	0.09		L(D); O(D); R(D); R/S(D)
Pepper (Capsicum annuum)[156]	0.12	3 h day^{-1}; 34 days	L(I);[d] S(N); O(D);[e] R(N)
	0.20		L(D);[d] S(N); R(D);[e] R(N)
White clover (Trifolium repens)[23]	0.05	4 h day^{-1}; 6 days	L(D); S(D); R(D); R/S(D)
	0.10		L(N); S(D); R(D); R/S(D)
	0.15		L(D); S(D); R(D); R/S(D)
Cotton (Gossypium hirsutum)[157]	0.25	6 h day^{-1}; 30-40 days	L(D); S(D); R(D); R/S(D)
Hybrid poplar (Populus deltoides x trichocarpa)[158]	0.050	5.5 h day^{-1}; 70 days	L(N); O(N);[f] R(N); R/S(N)
	0.085		L(D); O(N);[f] R(D); R/S(N)
	0.125		L(D); O(D);[f] R(D); R/S(N)
Parsley (Petroselinum crispum)[159]	0.20	4 h day^{-1}; 16 days	L(N); O(D);[c] R(D); R/S(D)

[1]Symbols: Leaf (L); shoot (S); other (O); root (R); root/shoot (R/S)
          No effect (N); increase (I); decrease (D)

[a]tuber yield   [d]weight per leaf

[b]tiller weight   [e]weight per fruit

[c]total plant weight   [f]stem weight

Sulfur Dioxide Effects

Early research with SO_2 effects on growth and yield of plants indicated that reductions ordinarily did not occur unless the SO_2 concentrations were quite high (e.g., >1.0 $\mu L\ L^{-1}$) and caused substantial visible foliar injury.[27,28,29]

However, in recent years numerous studies have indicated
that reductions in yield are not always dependent on
severe tissue destruction that is associated with very
high SO_2 concentrations, but may be caused by long-term
exposure to levels of SO_2 that may occur regionally in
some areas. Several books and reviews have given a
thorough treatment to these studies.[5,8,10,30] Table 3
and 4 summarize a few of these studies.

Several productive approaches to research of SO_2
effects on plant can be highlighted. One approach has
been to perform the studies in the vicinity of an actual
SO_2 emission source. For example, Guderian and Stratmann[31]
followed the growth of 12 crops grown at six sites located
at intervals (325-6000 m) from an iron-ore smelter in
Germany (Table 3). With this approach plants were exposed
to periodic episodes of SO_2 in which the concentrations
became lower and the frequency of exposure less with
increasing distance from the source. In this study, winter
wheat and potatoes (Solanum tuberosum) were fairly sensitive
and yields were reduced by about 10% at a site with a mean
concentration of 0.18 $\mu L\ L^{-1}$ during fumigation periods
(10-13% of the growing season). The yields of crops such
as oats (Avena sativa), spring wheat, clover and beets
(Beta vulgaris) were reduced 7-17% at a site having a mean
SO_2 concentration of 0.24 $\mu L\ L^{-1}$ (18-23% of the growing
season). One should keep in mind that SO_2 concentrations
were frequently greater than or less than the reported
means due to the fluctuation in concentrations that occur
near point sources. Thus, investigators were unable to
determine the relative importance of peak in relation to
mean concentration.

Since SO_2 is often a "point source pollutant", many
studies have emphasized research with recurrent short-term
exposures. Sprugel et al.[32] created their own SO_2 sources
by periodically releasing SO_2 into field plots from a pipe
system, a technique described in Millet et al.[33] Two years
of studies with soybean indicated that mean SO_2 concentrations of 0.09 to 0.12 $\mu L\ L^{-1}$ during exposure (8-10% of the
growing season) reduced soybean yields by 5-12%. Visible
injury to leaf tissue was not apparent after episodes, but
the plants did senesce earlier than those in control
(non-SO_2 treated) plots located nearby. As with studies
near actual point sources, the concentrations fluctuated
around the reported mean values. In contrast, Heagle

Table 3. Effects of sulfur dioxide (SO_2) on plant growth
and yield in field studies.

Species	Study conditions	SO_2 (μLL^{-1}) during study	during fumigation	Response[a]
12 Crops[31]	Plots located between 325 to 6000 m from SO_2 source in W. Germany; 2 yr study	0.02	0.15	N
		0.03	0.18	YR in winter wheat and potato
		0.05	0.24	YR in 8 crops
		0.08	0.34	YR in 10 crops
		0.14	0.45	9-70% yield loss for all crops
9 Trees and 2 bushes[31]	Plots located between 325 to 6000 m from SO_2 source in W. Germany; 2 yr study	0.02	0.15	Loss on 1 fruit bush
		0.03	0.18	> 10% loss in 8 species
		0.05	0.24	> 25% loss
		0.08	0.34	> 35% loss
		0.14	0.45	Severe damage
White Pine[160] (Pinus strobus)	Survey at 42 sites for mortality and volume increment at sites near SO_2 source in Canada	0.004-0.005	–	1.0% AM
		0.008-0.019	–	1.3% AM
		0.034-0.045	–	2.6% AM
Alfalfa[161] (Medicago sativa)	Field chambers; continuous; 68 days	–	0.06	26% YR
Soybean[34] (Glycine max)	Field chambers; 6 h day^{-1}, 133 days	–	0.10	N on seed yield
Soybean[32] (Glycine max)	Open-air fumigation; 76 to 113 h	–	0.09	6% YR
		–	0.10	5% YR
		–	0.12	12% YR
		–	0.19	12% YR
		–	0.25	19% YR
		–	0.30	21% YR
		–	0.36	16% YR
		–	0.79	45% YR
Soybean[35] (Glycine max)	Open-top chambers; 4 hr day^{-1}, 7 day wk^{-1}; growing season	–	0.03	2% YR
		–	0.08	10%
		–	0.37	32%
Soybean[47] (Glycine max)	Open-top chambers; 4 hr day^{-1}, 3 day wk^{-1}, 75 days	–	0.09	11-13% YR
		–	0.30	12-21% YR
		–	0.56	32-34% YR
Corn[36] (Zea mays)	Open-air fumigation; 36 h	–	0.07-0.67	N on yield

[a] AM, annual mortality; YR, yield reduction; N, no effect.

et al.[34] used field chambers in a study of SO_2 effects on
soybeans and found no effect on yield of seed after over
700 hours of treatment at 0.10 μL L^{-1} SO_2. One difference
from the work of Miller et al.[33] and Sprugel et al.[32] was
the absence of peak concentrations when chambers are used.

Heagle et al.[35] found in a later study with open-top
chambers that soybean yields were reduced at SO_2 concentrations similar to those used by Sprugel et al. (1980). As
is the case for O_3, variation in sensitivity to SO_2 occurs
among species. For example, in a study with field corn,
Miller et al.[36] found that 36 hours of exposure to SO_2
concentrations as high as 0.67 μL L^{-1} did not reduce
yields. When compared to the work with soybeans[32,34] it is
evident that there is a wide variation in species sensitivity to SO_2.

A common experimental approach for SO_2 studies has been
to expose plants to long-term fumigations in chambers
located in the greenhouse or in controlled environment
chambers. This approach has been used when the focus of the
study was on constantly elevated SO_2 levels occurring from
widespread diffuse sources or a high concentration of point
sources. In recent years this has been of primary concern
in Great Britain and in Europe, although a few areas in the
U.S. may experience similar conditions. Numerous studies
of this type have been reported, and some are highlighted
in Table 4. Roberts[30] has published an excellent, in-depth
evaluation of studies of this type and his conclusions are
as follows:

SO_2 $(\mu L$ $L^{-1})$

0.076-0.150   "for 1-3 months generally
                            produces significant yield
                            losses in the limited number
                            of crops studied;

0.038-0.076   for several months has
                            produced yield losses in
                            some studies but not all;
                            and

0.019-0.038   for several months has
                            produced beneficial as well
                            as detrimental effects on
                            yield."

It should be kept in mind that exceptions to Roberts'
conclusions are numerous, but considering the range in
species sensitivity and the diversity of experimental
approaches used, these are reasonable generalizations.

Table 4. Effects of sulfur dioxide (SO_2) on plant growth in greenhouse and controlled environment studies.

Species	SO_2 (μLL^{-1})	Exposure Duration	Response[a]
Ryegrass[162] (Lolium perenne)	0.07 0.12	continuous, 2-6 wks continuous, 9 wks	51% D in leaf d.w. 46% D in leaf d.w.
Hybrid Poplar[163] (Populus deltoides x P. trichocarpa)	0.25	8 hr day^{-1}, 5 day wk^{-1}, 30 days	D in growth of basal and terminal cuttings
Tobacco[46] (Nicotiana tabacum)	0.05	8 hr day^{-1}, 5 day wk^{-1}, 4 wks	D in dry weight of leaf, stem and root
Ryegrass[164] (Lolium perenne)	0.02	continuous, 87 days	66% I in shoot d.w., with low soil S; no effect with high soil S
Ryegrass[165] (Lolium perenne)	0.02- 0.20	continuous, 42- 194 days	I or D in biomass depending on conditions; conc. as low as 0.02 occasionally produced reductions
Wheat[166] (Triticum aestivum) (7 cv)	0.2 0.4 0.6	100 h	+ 15% plant d.w. - 6% plant d.w. - 30% plant d.w.
Ryegrass[167] (Lolium multiflorum)	0.068	continuous, 140 days	28% D in leaf d.w.
Timothy[167] (Phleum pratense)	0.068	continuous, 140 days	25% D in leaf d.w.
Soybean[168] (Glycine max)	0.25	4 h day^{-1}, 3 day wk^{-1}, 11 wks	17% D in plant d.w.

[a] D, decrease; I, increase

Table 5. Effect of sulfur dioxide (SO_2) on partitioning of biomass in several plant species.

Species	SO_2 (μLL^{-1})	Exposure Duration	Leaf	Shoot	Other	Root	R/S
Timothy[41] (Phleum pratense)	0.06 0.11	3 wks 6 wks	- -	N N	- -	D D	D D
Red pine[196] (Pinus resinosa)	0.20	91 h over 6 wks	-	I	-	D	D
Soybean[32] (Glycine max)	0.09- 0.36	76-113 h over growing season	D	D	D[a]	-	-
Fescue[38] (Festuca arundinacea)	0.1	6 h day$^-$; 12 days	-	N	-	D	D
Oat[37] (Avena sativa)	0.4	3 h day$^-$; 4 days	-	N	-	D	D

Symbols: No effect (N); increase (I); decrease (D); no data (-)

[a] harvest ratio

As with O_3, SO_2 may affect partitioning of biomass
among plant parts, and root growth seems to be inhibited
more than shoot growth (Table 5). For example, Heck and
Dunning[37] found that short-term exposure (3 hr on 4 days)
of oat to 0.4 μL L^{-1} SO_2 reduced root growth by 12% but did
not affect top growth. Similarly, exposure of tall fescue
to 0.1 μL L^{-1} SO_2 once a week for 12 weeks did not affect
shoot dry weight but decreased root weight by 11%.[38] Recent
studies have indicated that carbohydrate translocation may
be especially susceptible to exposure to SO_2. Noyes[39]
found that the translocation of ^{14}C from $^{14}CO_2$-labeled
bean leaves was reduced 39, 44, or 66% by a 2-h exposure
to SO_2 at 0.1, 1.0 or 3.0 μL L^{-1}, respectively. Teh and
Swanson[40] reported a 45% decrease in ^{14}C-assimilate translocation in bean exposed to 2.9 μL L^{-1} SO_2 for 2 h while
Jones and Mansfield[41] found both reduced root growth and
^{14}C translocation to roots of timothy (Phleum pratense)
exposed to 0.06 ppm SO_2 for six weeks. In these studies,
the reduced translocation to roots was attributed to
reduced phloem loading. Controlled field studies by
Milchunas et al.[42] showed that translocation of ^{14}C-
photosynthate from source leaves to developing leaves of
bluestem (Agropyron smithii) was stimulated by 12% at a
mean seasonal exposure level of 0.08 μL L^{-1}, although
translocation to roots was not markedly affected. Experiments with bean also demonstrated that SO_2 stress caused
significant alteration of translocation and partitioning
of photosynthate between plant parts, including developing
pods.[25]

Combined Ozone and Sulfur Dioxide Effects

Elevated concentrations of O_3 and SO_2 may sometimes
occur together in the ambient environment and a few studies
have been designed to test their combined effects on growth
and yield. Reviews by Reinert,[43] Ormrod,[44] and Heck et
al.[11] provide a more thorough discussion. Following the
early findings of Menser and Heggestad[45] that O_3 and SO_2
caused a synergistic injury response in tobacco (Nicotiana
tabacum), there has been concern whether plants would
react differently to pollutant combinations than what
would be predicted from studies with the single pollutants.
Thus far, no general patterns have emerged from these
studies as synergistic, antagonistic and additive effects
all have been found. The results seem to depend on the
species, the pollutant concentrations, the growth parameters

measured, and the conditions under which the experiments
were performed. Several experiments with soybean
illustrate this quite well. Tingey and Reinert[46] found a
synergistic inhibition of foliar and root growth by O_3 and
SO_2, while Heagle et al.[34] and Kress et al.[47] found no
statistically significant interaction of the two gases on
soybean yield. In a later study with soybeans, Heagle
et al.[35] found no interactive effects on yield at near-
ambient concentrations and an antagonistic effect at
higher concentrations. As a whole, these studies with
soybean illustrate the variety of responses that may be
obtained with one species.

Other species have been studied although perhaps not
as thoroughly as soybeans. For instance, additive effects
have been found with radish (Raphanus sativus)[48] and
American elm (Ulmus americana),[49] synergistic effects with
snap beans,[50] and antagonistic effects with alfalfa
(Medicago sativa)[46] and white bean.[51] Clearly, more
research is needed in order to understand the complex
interactive effects of air pollutants such as O_3 and SO_2.

PHYSIOLOGICAL EFFECTS

Photosynthesis and Ozone

About 90% of the dry weight of plants is derived from
photosynthetic CO_2 fixation. Processes related to CO_2
fixation should be sensitive indicators of air pollutant
stress, and should be included in studies designed to
understand the mechanisms of the stress response. One
concern in relating photosynthesis to plant productivity
is that plant growth is the result of carbon assimilation
and metabolism over its entire life-span, while measurements of photosynthesis or detailed studies of photosynthetic mechanisms are usually short-term. Still, even with
this limitation, studies of photosynthesis are one
important part of understanding air pollutant stress.

Reports have been published since the late 1950s
indicating that O_3 reduces photosynthetic rates of a wide
range of plant species. Much of the early work used quite
high O_3 concentrations due to its relatively unknown degree
of phytotoxicity and because of the lack of available data
on typical environmental concentrations, other than in

extremely impacted areas such as the Los Angeles Basin.
Many studies have been geared to address those relatively
acute exposures. In the interest of brevity, more recent
work using chronic exposures that are more typical of
regional O_3 stress will be discussed below.

Only a few studies have evaluated photosynthesis
under conditions of ambient O_3 stress in the field. One
such study with ponderosa pine in the San Bernardino
Forest[52,53] was performed by Coyne and Bingham. This area
is heavily impacted by O_3 pollution, and concentrations as
high as 0.6 $\mu L\ L^{-1}\ O_3$ have been recorded. Eighteen-year-
old saplings were separated into three classes based on O_3
injury symptoms, and photosynthesis of individual fascicles
of three age classes were measured monthly from May to
October. Since O_3 was not excluded from any of the trees,
comparisons with controls grown under low O_3 conditions were
not possible. Their findings indicated that the decline in
photosynthesis and stomatal function, that is normally
associated with aging, was accelerated as O_3 injury
symptoms increased. Furthermore, considerable ecotypic
variation occurred among individual trees. When photosynthesis was reduced to about 10% of the maximum rates,
needle abscission occurred. This typically happened after
cumulative doses of 450 to 800 $\mu L\ L^{-1}h$. The less sensitive
trees seemed to require the higher doses before needles
were dropped. Interestingly, stomatal conductance in
current-year needles was highest in the more severely
injured trees indicating that stomatal function may play
a role in the observed differential sensitivity of ponderosa
pine. In all cases, losses in photosynthetic capacities
exceeded reductions in stomatal conductance, suggesting
that chloroplast function rather than stomatal function was
primarily responsible for the loss in photosynthetic
capacity.

Photosynthetic responses to O_3 stress have more
commonly been measured under controlled experimental
conditions, and several studies using controlled O_3 exposures
have illustrated that O_3 reduces photosynthesis at concentrations at or near common ambient concentrations. In
Figure 1, results of a few representative experiments are
shown. While some species [e.g., loblolly pine (Pinus
taeda) and one O_3-insensitive clone of white pine (Pinus
strobus)] showed little photosynthetic depression, some
species showed very significant reductions at O_3 concentra-

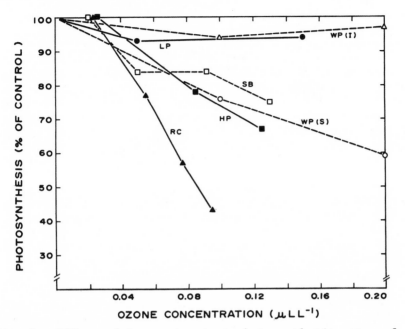

Fig. 1.   Effect of O3 treatment on photosynthetic rates of
five species.
WP -- Pinus strobus, white pine, insensitive (I) or
         sensitive (S) clone, Yang et al.;[54] 4 h day^{-1}, 50 days
LP -- Pinus taeda, loblolly pine, Barnes;[68] 84 days,
         continuous
SB -- Glycine max, soybean, Reich et al.;[61] 6.8 h day^{-1},
         28 days
HP -- Populus deltoides x P. trichocarpa, hybrid poplar,
         Reich;[60] 5.5 h day^{-1}, 62 days
RC -- Trifolium repens, red clover, Reich and Amundson;[55]
         10 h day^{-1}, 18 days

tions which commonly occur. The data with white pine is
especially interesting since such a great difference was
noted between clones.   Reich and Amundson presented a
summary of their research on seven crop and tree species
that were exposed to chronic O3 stress under laboratory
and/or field conditions. Some of these data are shown
in Figure 1 [soybean, red clover (Trifolium pratense), and
hybrid poplar (Populus deltoides x P. trichocarpa)]. Their
data generally indicate a linear decline in photosynthetic

rates of individual leaves with increasing O_3 concentrations, although very little effect was found on photosynthetic rates of white pine, sugar maple (<u>Acer saccharinum</u>) or red oak (<u>Quercus rubra</u>) under the study conditions (data not shown). (Note that some clones of white pine are sensitive, Fig. 1.) It is difficult to compare relative species sensitivity in these studies since exposure durations (hours per day and days per experiment) and exposure conditions differed. However, after converting to the total treatment dose ($\mu L \ L^{-1}h$) during exposure periods, Reich and Amundson[55] made a tentative comparison of their data. Red clover, wheat, soybean, and hybrid poplar appeared to be the most sensitive to photosynthetic depression while sugar maple also appeared sensitive when the data were presented in this fashion. White pine and red oak were much less sensitive. It is pertinent that the four most sensitive species also had the highest inherent rates of photosynthesis. It is likely that the higher gas exchange rates in these species resulted in greater O_3 flux into the leaf, and this may be partially responsible for the greater apparent sensitivity of these species. It is important to note that the low and intermediate O_3 concentrations used in the studies are quite comparable to growing season concentrations found in much of North America and elsewhere. In the eastern and midwestern U.S., seasonal O_3 doses of 30 to 50 $\mu L \ L^{-1}h$ are common during daylight hours over an average growing season. Doses as low as 10 $\mu L \ L^{-1}h$ caused substantial photosynthetic depression in leaves of five of the seven species in these studies.

Other studies have indicated less effect of O_3 on photosynthesis at near-ambient concentrations, although these differences may be generally attributed to shorter histories of exposure. Black et al.[56] found that 0.063 $\mu L \ L^{-1} \ O_3$ caused little apparent effect on photosynthesis of 3-week-old broad bean (<u>Vicia faba</u>) during a 4-h exposure. A concentration of 0.151 $\mu L \ L^{-1}$ did reduce photosynthesis by about 35% but the rate had recovered to nearly its initial level the following day. With concentrations of 0.217 or 0.290 $\mu L \ L^{-1} \ O_3$, recovery was not apparent indicating permanent destruction of the photosynthetic tissue had occurred. In a study using soybean cv 'Mcall' at the R5 stage of development, Le Sueur-Brymer and Ormrod[57] found that 0.067 $\mu L \ L^{-1} \ O_3$ for 7.5 h day^{-1} for 5 days reduced photosynthetic rates of the entire plant from about

2 to 18%, depending on when the measurements were taken.
During the last 2 days of treatment the effects were
generally less pronounced, indicating a pattern of recovery
or development of tolerance to the stress. In another
study, a 2-h treatment of soybean cv 'Essex' with 0.20
μL L^{-1} O$_3$ caused only very slight reductions in photosynthesis, which the authors found to be statistically
nonsignificant.[58] Pell and Brennan[59] found that a 3-h
treatment of bean with 0.25 to 0.30 μL L^{-1} O$_3$ resulted in
a 22% reduction in photosynthesis immediately following
treatment but that rates had recovered to the control level
within 21 h.

In most studies, measurements are made for a limited
time after O$_3$ exposure and thus do not represent the
complete photosynthetic capacity of the leaves throughout
their development. However, Reich[60] and Reich et al.[61]
did measure photosynthesis of hybrid poplar and soybean
during leaf development. These studies indicated that
chronic O$_3$ exposure reduced the capacity for photosynthesis
throughout the developmental period of the leaves for both
species. This was due in part to reductions in leaf
chlorophyll and acceleration of senescence. With hybrid
poplar, O$_3$ also tended to lower light saturation maxima
and to reduce apparent quantum yield while elevating light
compensation points. Coyne and Bingham[62] also measured
light saturation relationships for photosynthesis and
diffusive conductance of young but fully-developed leaves
of bean that had been exposed to 0.072 μL L^{-1} O$_3$ for 66 h
over a period of approximately 18 days. The maximum
photosynthetic rate at light saturation was reduced 18%
by the O$_3$ treatment. Thus, O$_3$ does seem to reduce
maximum photosynthetic efficiency under conditions of
chronic exposure.

Several conclusions concerning O$_3$ effects on photosynthesis may be drawn from the research done to date.
Short-term exposures that do not cause visible injury may
reduce photosynthesis, but the plants may recover within
hours. Higher O$_3$ concentrations may cause tissue destruction that permanently reduces the capacity for photosynthesis. Most importantly, chronic exposures at lower O$_3$
levels can reduce the ability of the plant to photosynthe-
size, probably by accelerating senescence. The latter case
is what is usually encountered in areas impacted by O$_3$.

Respiration and Ozone

Effects of O_3 on respiration rates may be due to injury, the action of O_3 on other processes, or effects on respiratory processes themselves. Not surprisingly, research has shown that O_3 will affect respiration if injury occurs. Todd[63] determined that rates of respiration in pinto beans were stimulated by O_3 only when visible injury occurred, but that the greater the stimulation, the more complete was recovery. Similarly, the increase in respiration in leaf discs of bean after 24 h exposure to O_3 led Pell and Brennan[59] to propose that enhanced respiration was a consequence of cellular injury. Stimulated respiration rates and uncoupling of oxidative phosphorylation were also reported by Macdowall and Ludwig[64] after the appearance of visible damage. While Macdowall[65] did later report that larger doses of O_3 resulted in increased respiration and visible injury in tobacco, low concentrations of O_3 not causing visible injury resulted in an inhibition of respiration and a reduction in mitochondrial phosphorylation.

There is also evidence that exposure to low concentrations of O_3 can result in the stimulation of respiration in the absence of visible injury. For example, Todd and Propst[66] found that O_3 caused a two- or three-fold stimulation in the respiration of coleus (Coleus, spp.) and tomato (Lysopersicum esculentum) in the absence of visible injury, while Dugger and Ting[67] observed that one of the first noticeable changes after the start of an O_3 exposure period was an increase in respiration rates. Barnes[68] found that respiration was stimulated when pine seedlings were exposed to 0.05-0.15 μL L^{-1} O_3. This led Barnes to suggest that a stimulation in respiration, along with the observed inhibition of photosynthesis, would probably be a drain on carbohydrate supply resulting in a reduction in growth and vigor.

There is some evidence that the respiratory function of tissues other than leaves may be affected by O_3. For example, Hofstra et al.[69] reported that the respiratory activity of roots was very sensitive to the changes induced in the leaves of bean exposed to O_3. Within 24 h of the beginning of exposure to 0.15 ppm O_3, root growth was inhibited and the CO_2 evolution of the root system was reduced. Inhibition of root activity occurred before the appearance of visible foliar injury and before any reduction in photosynthetic area. However, it is likely that the effects in the roots were caused by reduced photosynthesis resulting in reduced carbohydrate supplies transported to the roots. One piece of evidence that respiratory responses may be directly affected was found by Anderson and Taylor.[70] They reported increased carbon dioxide evolution in non-photosynthetic tobacco callus exposed to O_3.

Ultrastructural changes in mitochondria due to O_3 have been found by Pell and Weissburger[71] while Lee[72] reported increased permeability of mitochondrial membranes. Such ultrastructural changes are likely to alter the components of the electron transport system, and changes in ATP levels have been observed.[73] Lee[74] reported that uncoupling of oxidative phosphorylation in mitochondria occurred when tobacco leaves were exposed to O_3. These changes in phosphorylation were observed before changes in respiration occurred.

Studies of O_3 effects on photorespiration are few despite the importance of this process in C-3 plants. The reasons for this probably are related to the difficulties in making quantitative measurements. Several studies have looked at pool sizes of metabolites of the glycolate pathway. Ito et al.[75] found pool sizes of glycine and serine (and $13C$ incorporation into them) in bean were increased by O_3 exposure which they interpreted as evidence that carbon flow through the pathway was stimulated. Johnson[76] found a similar buildup of glycine and serine with exposure of while clover to O_3. However, it seems that another possible interpretation of these results would be that the metabolism or translocation of the two amino acids was blocked.

Thus, O_3 may modify the respiratory activity of plant tissues and either stimulation or inhibition may occur. While the rate of respiration, especially dark respiration, is less than photosynthesis, it is still an important part of the energy balance of the plant. It is likely that much of the O_3 effect on respiration is due to alteration of membrane permeability, although acceleration of energy use for repair cannot be ruled out.

Cellular Mechanisms and Ozone

Ozone is a strong oxidant and is extremely reactive with a variety of substances found within living systems. Therefore, it is unlikely that it passes very far into the cell before reacting; instead its effect may occur initially at either the cell wall or the plasmalemma. In aqueous solutions O_3 decomposes to form molecular oxygen and reactive products such as hydroxyl ions, various radicals, and hydrogen peroxide. Ozone reacts with biologically important compounds including unsaturated fatty acids and ring-containing compounds.[77] Ozone will break down the nicotinamide rings of NAD(P)H[78] and also will oxidize some aromatic amino acids.[79] Generally, free sulfhydryls are oxidized to disulfides. Such reactions undoubtedly lead to many of the biological effects of O_3.

Unicellular algae have been used as models to study the primary site of O_3 interaction with plant cells.[16] Treatment of a culture of Chlorella cells with O_3 caused an immediate increase in efflux of potassium ions suggesting a two- to three-fold increase in the passive permeability of the membrane to potassium.[80] Sutton and Ting[81] found that leaves showed an increased permeability to 2-deoxyglucose after O_3 exposure. Permeability returned to normal after several days and the reversal appeared to require energy.

The effect of O_3 on isolated chloroplasts has also been studied. For example, Coulson and Heath[82] found that ozone bubbled into a suspension of isolated spinach chloroplasts inhibited electron transport in both photosystems I and II without uncoupling ATP production. The production of ATP declined concurrently with both electron transport and amine-induced swelling (an indication of the H^+ gradient). Thus, the same net increase in permeability to ions seen for the plasma membrane might also occur in grana membranes treated with O_3.

Using isolated plastids with intact envelopes, Nobel and Wang[83] observed alterations in membrane permeability to several neutral compounds such as glycerol and erythritol. The increase in permeability is not general, however, because intact plastids remained impermeable to sucrose even at the highest O_3 dose. Permeability increases in chloroplasts could lead to a loss of intermediates of the

Calvin cycle and slowing of the rate of CO_2 fixation and photophosphorylation. This might explain the O_3 inhibition of bicarbonate-stimulated oxygen evolution observed earlier by Coulson and Heath.[82]

Since O_3 is a strong oxidant, it is not surprising that it often affects enzyme activity, especially those that have active sulfhydryl groups. Several enzymes involved in the metabolism of structural and non-structural carbohydrates are affected by O_3. Ordin and Hall[84] and Ordin, Hall and Kindinger[85] found that O_3 partially inhibited DGPG- and UDPG-β-glucan glucosyltransferases indicating a possible interference with cellulose biosynthesis. Dass and Weaver[86] found that cellulase activity increased in bean following O_3 treatment. Starch hydrolysis also may be modified by relatively low concentrations of O_3. For example, 0.05 μL L^{-1} O_3 for 2-6 h inhibited starch hydrolysis in cucumber (Cucumis sativus), bean and other species.[87] They suggested that O_3 may have inhibited amylase or starch phosphorylase.

Other pathways of carbon flow also may be modified by O_3. Tingey et al.[88] found that treatment of soybeans with a rather high concentration of O_3 (0.5 μL L^{-1}) for 2 h increased the activity of glucose-6-phosphate dehydrogenase and decreased glyceraldehyde-3-phosphate dehydrogenase activity. This implies an increase in the activity of the pentose phosphate pathway and a reduction in glycolysis, a fairly common response to stress. In this same study, the activities of polyphenol oxidase, peroxidase and phenylalanine ammonia lyase were depressed immediately following the O_3 exposure, but then recovered and exceeded the activities from control plants, indicating that phenol metabolism was stimulated. This is probably the result of lesion formation. Peroxidase activity has been found to be elevated due to O_3 treatment by several workers.[86,89]

Lipid metabolism is also affected by O_3. Mudd et al.[90] found an inhibition of glycolipid biosynthesis in spinach (Spinacea oleracea) chloroplasts. Tomlinson and Rich[91] found more sterol glycosides and less free sterol in bean and spinach leaves treated with a high O_3 concentration (0.5 μL L^{-1} for 1 h). In contrast, long-term chronic treatment of soybean with lower O_3 concentrations caused an increase in free sterols in the leaf while steryl ester and steryl glycosides decreased.[92] Although it is difficult

to generalize from these few observations, the latter
study probably reflects a more environmentally realistic
case.

The research on mechanisms of O_3 effects seems to
suggest that the primary site of O_3 toxicity is the
plasmalemma. Any disturbance of this membrane can lead to
a significant disruption of metabolism, and a number of
metabolic alterations due to O_3 have been documented. It
is uncertain whether significant amounts of O_3 or its
reaction products penetrate into the cell and directly
affect other cellular organellles or processes, or whether
all metabolic changes are indirect. Mudd[93] does present
some evidence that O_3 may move into cells and directly
alter metabolism.

Photosynthesis and Sulfur Dioxide

The inhibitory effects of SO_2 on photosynthesis have
been documented in several reviews.[1,16,94,95,96] In spite
of the considerable information available, it is usually
not possible to predict the effect of SO_2 on CO_2 assimilation. The majority of investigations indicate that SO_2
exposure depresses net photosynthesis, although some
workers report temporary enhancement.[97,98] Sometimes these
enhanced rates can be attributed to increased stomatal
conductance, or possibly to depressed photorespiration,
but many of the reported reductions probably do not result
entirely from changes in these factors. Depressions of
photosynthetic rates may occur within minutes to hours
after the start of exposure, are often reversible and may
not be accompanied by major visible injury, at least at
low concentrations.[97] At higher concentrations, responses
often are not reversible and often are associated with the
appearance of visible injury.

Species vary in the degree of photosynthetic response
to SO_2, as illustrated in Figure 2, which was constructed
from the data of several workers. In a few cases,
inhibition of photosynthesis was seen at low SO_2 concentrations (<0.1 μL L^{-1}). For example, broad bean was
especially sensitive at very low concentrations with a 7%
reduction at 0.035 μL L^{-1} for 7 h.[97] In this study, the
photosynthetic rates recovered to control values within
an hour after termination of exposure, and visible injury
did not occur (even at concentrations up to 0.175 μL L^{-1}).

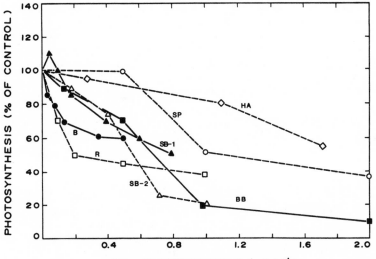

Fig. 2. Effects of SO_2 treatment on photosynthetic rates
of six species.
 SP -- <u>Pinus</u> <u>sylvestris</u>, Scots pine, Oleksyn;[133] 6 h day^{-1},
      2 days
  B -- <u>Vicia</u> <u>faba</u>, broad bean, Black and Unsworth;[97] 7 h
SB-1 -- <u>Glycine</u> <u>max</u>, soybean, Muller <u>et</u> <u>al</u>.;[134] 4 h day^{-1}
      average, approximately 6 days
SB-2 -- <u>Glycine</u> <u>max</u>, soybean, Carlson;[135] 2 h
  R -- <u>Oryza</u> <u>sativa</u>, rice, Katase <u>et</u> <u>al</u>.;[136] 5 h
 BB -- <u>Phaseolus</u> <u>vulgaris</u>, bean, Taylor and Selvidge;[137]
      6 h
 HA -- <u>Heteromeles</u> <u>arbutifolia</u>, Winner and Mooney;[98] 8 h

In other studies with a variety of species, much higher
concentrations have been required to cause significant
inhibition of photosynthesis [e.g., Scots pine (<u>Pinus</u>
<u>sylvestris</u>) in Fig. 2]. Occasionally, a stimulation at low
SO_2 concentrations has been seen (soybean in Fig. 2). A
variety of explanations for stimulation have been offered
such as increased sulfur nutrition, effects on stomatal gas
exchange and stimulation of energy requiring repair
processes. The reasons for the apparent variation in
response among species is uncertain. Environmental factors
during plant growth and during the exposure period most

certainly are involved to some extent. For example, interactive effects of relative humidity (RH) and SO_2 on yield
and physiological processes have been observed. Effects of
SO_2 are generally greater when RH is high, probably because
stomatal conductances are greater than when RH is low.[99]

Photosynthesis and Combined Ozone and Sulfur Dioxide

The reported synergistic interaction of O_3 and SO_2
on growth has prompted investigators to look to photosynthesis for a possible explanation. For example, Ormrod
et al.[102] studied the response of broad bean to O_3 with
or without SO_2 added at a concentration of 0.04 $\mu L\ L^{-1}$ for
4 h. The results indicate a synergistic response to
concentrations of O_3 below 0.11 $\mu L\ L^{-1}$ and an additive or
antagonistic response to O_3 concentrations above 0.11
$\mu L\ L^{-1}$. Other plants seem to be less sensitive to low
concentrations of O_3 and SO_2 combined, although higher
concentrations may cause synergistic effects. For example,
Carlson[100] found that maple (Acer saccharum) and ash
(Fraxinus americana) showed a synergistic depression with
0.5 $\mu L\ L^{-1}$ concentrations of SO_2 and O_3 for one week,
although oak (Quercus velutina) did not. Similarly,
Furukawa and Totsuka[101] found a synergistic decrease with
sunflower (Helianthus annuus) at concentrations of 0.20
$\mu L\ L^{-1}$L each of SO_2 and O_3. With soybean, Le Sueur-Brymer
and Ormod[57] tested the effects of SO_2 (0.30 μLL^{-1}) and O_3
(0.067 $\mu L\ L^{-1}$) separately and combined. Each pollutant
separately tended to reduce photosynthesis during 5 days
of fumigation and the combination caused a further decline,
but the interactive effect of the two gases was not
statistically significant. In another study with soybeans,
Chevone and Yang[58] found that either 0.20 $\mu L\ L^{-1}$ O_3 or 0.70
$\mu L\ L^{-1}$ SO_2 administered separately for 2 h did not significantly affect photosynthetic rates, but the combination of
the 2 gases at the same concenteations caused a 70%
reduction.

Currently, no research has conclusively tied synergistic reductions in photosynthesis due to SO_2 and O_3 to
synergistic growth reductions, although correlative
information suggests this may contribute. This is not
surprising since both pollutants also modify so many other
processes within the plant. However, most studies of
photosynthetic response to O_3 and SO_2 mixtures are not
designed to critically test growth and yield reductions.

Respiration and Sulfur Dioxide

Respiratory responses to SO_2 are not well understood,
even though many workers have studied the effects of a
wide range of concentrations on higher plants. Some
researchers have found little effect on respiration from
SO_2. For example, Thomas and Hill[103] found no effects on
dark respiration in plants exposed to 1 μL L^{-1} SO_2 for 1 h.
A similar lack of response to high concentrations has also
been reported by Katz,[104] Sij and Swanson[105] and Furukawa
et al.[106] However, the equipment and chambers used in
these studies may not have been adequate to measure the
rather small changes in CO_2 or O_2 concentrations necessary
in measurements of respiration. The majority of recent
research using proper chambers and equipment does indicate
some effect of SO_2 on respiratory processes, and both
inhibition[107] and stimulation[97,108] of dark respiration
has been reported.

Other types of evidence for SO_2 effects on the respiratory capacity of plants has been obtained. For instance,
ultrastructural changes in the mitochondria of lodgepole
pine (Pinus contorta) were observed[109] and an inhibition
of ATP formation and phosphorylating activity of mitochondria has been found observed in plants exposed to
SO_2.[110-112] As a whole, the evidence suggests that SO_2
may either stimulate or inhibit respiration, depending on
the experimental conditions and the species; this may or
may not be accompanied by visible injury. Respiratory
rates may return to control values following exposure as
long as concentrations are relatively low. This may
reflect the capacity of plants to detoxify sulfite or
repair damage incurred by exposure to toxic concentrations.

Studies of the effects of SO_2 on photorespiration are
limited in number, although more has been done with SO_2
than O_3. Assessments of pollutant-induced changes in this
process are of particular importance since rates of photorespiration probably are significantly higher than those
of dark respiration.[113] Koziol and Jordan[114] reported an
increase in photorespiration of bean with increasing SO_2
concentrations. They attributed this increase (up to 150%)
to a greater use of energy in repair processes. Several
workers have reported apparent inhibitions of photorespiration. For example, Furukawa et al.[106] found that
photorespiratory rates of sunflower plants were inhibited

by exposure to 1.5 µL L^{-1} SO$_2$ for 30 min. Koziol and
Cowling[115] reported that when rye grass (Lolium perenne)
was exposed to SO$_2$, an increase in ^{14}CO$_2$ photoassimilation
was observed which they attributed to an inhibition of the
photorespiratory cycle. Libera et al.[116] demonstrated that
one of the photorespiratory intermediates, glycolic acid,
accumulated in leaves of spinach exposed to a high concentration of SO$_2$. This was attributed to a blocking of
glycolate oxidation. Spedding and Thomas[117] and Soldatini
and Ziegler[118] have shown that sulfite may inhibit
glycolate oxidase.

As is typical of the effects of pollutant stress on
many variables, the type or degree of response varies
depending on the species, the pollutant concentration, the
environmental conditions and the measurement technique.
Pollutant effects on respiration deserve far more study
because of the relationships among dark respiration,
photorespiration and photosynthesis in determining plant
productivity.

Cellular Mechanisms and Sulfur Dioxide

There are a variety of hypotheses and volumes of data
concerning SO$_2$ effects on metabolism.[1,5,6,16,94,95] The
following discussion will touch upon some of those
mechanisms. In aqueous solution, SO$_2$ becomes H$_2$SO$_3$ and at
about pH 5.4 its dissociation products, HSO$_3$$^{-1}$ (bisulfite)
and SO$_3$$^{-2}$ (sulfite) occur in equal proportions, with
sulfite being the dominant species at higher pH values.
Both HSO$_3$$^{-1}$ and SO$_3$$^{-2}$ can form addition products with
metabolites containing carbonyl functional groups such as
sugars, glyconic acids, aldehydes and organic acids. Also,
the absorption of SO$_2$ increases the acidity of aqueous
solutions and may cause localized alterations in pH. Thus,
the effects of SO$_2$ on metabolism can result from disturb-
ances of pH, from competition with substrate (e.g.,
HCO$_3$$^{-1}$) in enzymatic reactions, or from the alteration of
the normal substrate by forming addition products.

Guderian and van Haut[119] suggested that SO$_3$$^{-2}$ could
affect membrane integrity. Interference with the structure
and permeability of membranes and associated enzymes will
result in alterations in many biochemical processes in the
cell. Puckett et al.[120] suggested that cleavage of
disulfide linkages by SO$_3$$^{-2}$ was responsible for protein and

membrane disruption, which would explain the increased
potassium efflux following exposure to SO_2.[121] Malhotra[109]
found that ultrastructural changes in chloroplast
membranes induced by SO_2 were associated with depression
in Hill reaction activity. Alscher-Herman[122] has suggested
that a possible explanation for the effect of SO_3^{-2} may be
the competitive action for binding sites in chloroplast
thylakoid membranes. High concentrations of SO_2 also
caused the degradation of chlorophylls to phaeophytins,
which leads to senescence[123] and prolonged exposures to
low concentrations of SO_2 resulted in injury at the
molecular level by affecting enzymes such as chloro-
phyllase.[124]

The HSO_3^{-1} addition product of glyoxylate (glyoxylate
bisulfite) was isolated from rice leaves exposed to SO_2[125]
and was shown to be an effective inhibitor of glycolate
oxidase.[126,127] Inhibition of glycolate oxidase (or other
enzymes in the photorespiratory pathway) is likely to be
detrimental to the plant in the long term since this may
divert so much carbon from the C_3 cycle that insufficient
ribulose-1,5-bisphosphate (RuBP) is generated to maintain
normal rates of photosynthesis. Glyoxylate bisulfite was
also found to inhibit the activities phosphoenolpyruvate
carboxylase and malate dehydrogenase, two enzymes involved
in the initial steps of C_4 photosynthesis,[128] as did
SO_3^{-2}.[129]

In view of similarities between SO_2 and CO_2 and
their ionic forms, the influence of SO_2 on RuBP carboxylase
has been extensively studied. Ziegler[130] found that SO_3^{-2}
inhibited RuBP carboxylase competitively with respect to
HCO_3^{-1}. Presumably SO_3^{-2} or HSO_3^{-1} replaces HCO_3^{-1} by
reacting at the same enzyme reaction site. Sulfite showed
a non-competitive inhibition with respect to RuBP and
Mg^{2+}. However, the type of effect on photosynthesis by
SO_3^{-2} is concentration-dependent. Libera et al.[116] found
that SO_3^{-2} concentrations less than 1 mM enhanced the fixation of HCO_3^{-1} by intact chloroplasts of spinach.
Gezelius and Hällgren[131] examined crude extracts of
spinach using the same assay conditions as Ziegler[130] and
found that SO_3^{-2} was a less potent inhibitor than claimed
previously and that the pattern of inhibition was non-
competitive. Part of the difference between these results
may be due to the conditions under which the enzyme was
assayed. Some recent work[132] suggested that the SO_3^{-2}

inhibition of RuBP carboxylase is very complex. Using a
fully activated state of the enzyme from wheat and
spinach, they also reported the kinetics were mixed and
changed with time. They found a progressive inactivation
by low concentrations of SO_3^{-2}, suggesting that the
potential effects of RuBP carboxylase may have been
underestimated.

The research conerning the physiological and
biochemical effects of SO_2 indicates that a number of
mechanisms probably are involved. As with O_3, SO_2 seems
to affect membrane structure and function. But unlike
O_3, SO_3^{-2} and HSO_3^{-1} or their conversion products may
exist for some time in the cell and have other effects
(e.g., direct inhibition of enzyme reactions). It seems
unlikely that any single mechanism accounts for the
effects of SO_2 on plant productivity.

Carbohydrate Pools and Combined Ozone and Sulfur Dioxide

A number of observations have been made concerning
O_3 and SO_2 effects on non-structural carbohydrate pools.
Some of these effects are summarized in Table 6. Increases,
decreases or no effects have been observed depending on the
particular study. Generalizations from the available data
are difficult. Carbohydrate pools are a reflection of a
number of interrelated processes within the plant. While
reductions in photosynthesis induced by O_3 and SO_2 would
seem to suggest that soluble carbohydrate pools and starch
reserves would decrease, the possible inhibition of
respiratory reactions and effects on other facets of
carbohydrate metabolism may tend to counteract carbohydrate
depletion. Part of the difficulty in interpretation is due
to the lack of comprehensive data on any particular plant/
pollutant combination. Furthermore, the pattern of carbon
assimilation and allocation differs among species and among
different growth stages within a species. It is to be
expected that pollutant-stress effects on non-structural
carbohydrate pools would be expressed differently in these
different cases. Ozone and SO_2 stress also may affect
cellular water relations and influence osmoregulatory
processes which can cause redistribution of C among the
various metabolic pools. Since carbohydrate pools are
intimately involved in plant growth, it is clear that
further research is needed in this area to understand their
relationship to the air pollutant stress phenomena.

Table 6. Effects of ozone (O_3) and sulfur dioxide (SO_2) on carbohydrate pools.

Plant Species	Pollutant Treatment $\mu L\ L^{-1}$ (duration)	Effect
Ozone		
Five Pinus species[68]	0.05, 0.15 (5 wks)	Total SS, RS in primary needles(I)
Rough Lemon[170,171] (Citrus limon)	0.15-0.25 (8 h day^{-1}, 5 days wk^{-1}, 9 wks)	RS(I); S in leaves(D)
Ponderosa Pine[172] (Pinus ponderosa)	Ambient exposure in field	SS in phloem(D)
Ponderosa Pine[173] (Pinus ponderosa)	0.3 (9 h day^{-1}, 33 days)	SS(I); P(D)
Broccoli[174] (Brassica oleracea)	0.2 or 0.35 (6 h day^{-1}, 1.5 days wk^{-1}, 105 days)	TC(I)
Corn[174] (Zea mays)	0.2 or 0.35 (2.5 h day^{-1}, 3 days wk^{-1}, 71 days)	TC(D)
Tomato[174] (Lycopersicon esculentum)	0.2 or 0.35 (2.5 h day^{-1}, 3 days wk^{-1}, 116 days)	TC(D)
Soybean[175] (Glycine max)	0.5 (2 h)	RS, S(D)
Ponderosa Pine[26] (Pinus ponderosa)	0.08 (6 h day^{-1}, 20 wks)	SS, S in roots(D); SS, S in tops(I)
White Pine[176] (Pinus strobus, Loblolly Pine, P. taeda)	0.1, 0.2, 0.3, 0.6, 1.0 (7 or 21 days)	^{14}C in SS(D); ^{14}C in sugar phosphates(I)
Potato[177] (Solanum tuberosum)	0.15 or 0.20 (1 day every 2 wks for growing season)	SS, S (variable)
Clover[182] (Trifolium repens)	0.1 (6 h day^{-1}, 5 days)	TNC in shoot(D)
Sulfur Dioxide		
Norway Spruce[178] (Picea abies)	Ambient conditions	TS in injured needles(I)
American Elm[179] (Ulmus americana)	2 (6 h, assayed 24 h after exposure)	Non-structural carbohydrates in leaves, stems, and roots(D)
Ryegrass[180] (Lolium perenne)	0.019, 0.148 (29 days, harvested, then additional 22 days)	TC after 29 days(I); after additional 22 days(D); fructosans(D)
Bean[114] (Phaseolus vulgaris)	0.40, 0.77, 0.53, 3.06, 4.03 or 10.66 (for 24 h)	RS, S(I) up to 3.06 $\mu L\ L^{-1}$; (D) at higher conc. SO_2
Jack Pine[181] (Pinus banksiana)	0.34-0.51 (96 h)	RS(I); non-RS(D)

[1]Symbols: Increase (I); Decrease (D); Soluble Sugars (SS); Reducing Sugars (RS); Starch (S); Polysaccharide (P); Total Carbohydrates (TC); Total Sugars (TS).

CONCLUSIONS

This review has briefly highlighted research relating
to O_3 and SO_2 stress effects on vegetation over a range
of levels of biological organization and complexity (i.e.,
whole plant to biochemical mechanism). Even though
considerable progress has been made in each of these
areas, in most cases it is not possible to link these
biological responses in a predictive manner. That is, we
are unable to say with certainty that a specific perturba-
tion of an enzymatic reaction or physiological process by
O_3 or SO_2 will translate to any quantifiable effect on
growth or development. This is hardly surprising in view
of the complexity of the plant system and our general lack
of understanding of how, or to what degree, plants respond
to any stress. The problem is further compounded by
interactions of air pollutant stress with other environmental variables or stresses which modify plant growth.

There is one additional reason that the mechanisms of
air pollution stress are not well-coupled to the ultimate
effects on plant growth. Too often, an individual
researcher, or research program, emphasizes only a very
limited part of the total problem. This is understandable
due to limitations in manpower, funding, facilities, as
well as the interests and training of the individuals
involved. However, this often has led to isolation of the
component parts of the research to the extent that it is
impossible to integrate the pieces "after the fact".
Careful consideration should be given to the application
of the research in understanding the total problem,
namely, air pollution effects on plant productivity.

ACKNOWLEDGMENTS

I thank Drs. Allen Haegle, Walter Heck, and Steven
Shafer for their helpful reviews. Special thanks are
extended to Ms. Clara Edwards, Ms. Beth Godavarti, Miss
Ramona Lane, and Ms. Benita Perry for typing and other
aspects of manuscript preparation and to Ms. Jeanie
Hartman and Ms. Anne Bowman for library assistance.

REFERENCES

1. MUDD, J.B., T.T. KOZLOWSKI, eds. 1975. Responses of
   Plants to Air Pollutants. Academic Press, Inc.,
   New York, 383 pp.
2. ORMROD, D.P. 1978. Pollution in Horticulture.
   Elsevier Scientific Publishing Co., Amsterdam and
   New York, 260 pp.
3. PELL, E.J. 1979. How air pollutants induce disease.
   In Plant Disease. (J. Horsfall, E.B. Cowling, eds.),
   Vol. IV, Academic Press, Inc., New York, pp.
   273-292.
4. SMITH, W.H. 1981. Air Pollution and Forests:
   Interactions Between Air Contaminants and Forest
   Ecosystems. Springer-Verlag, New York, Heidelberg,
   Berlin, 379 pp.
5. UNSWORTH, M.H., D.P. ORMROD, eds. 1982. Effects of
   Gaseous Air Pollution in Agriculture and Horticulture.
   Butterworth Scientific, London, 522 pp.
6. KOZIOL, M.J., F.R. WHATLEY, eds. 1984. Gaseous Air
   Pollutants and Plant Metabolism. Butterworths,
   London, 446 pp.
7. TRESHOW, M., ed. 1984. Air Pollution and Plant Life.
   John Wiley and Sons Ltd., Chichester, England,
   486 pp.
8. GUDERIAN R. 1977. Air Pollution, Phytotoxicity of
   Acidic Gases and Its Significance in Air Pollution
   Control. Springer-Verlag, Berlin, 127 pp.
9. GUDERIAN, R., ed. 1985. Air Pollution by Photochemical Oxidants. Springer-Verlag, New York,
   346 pp.
10. WINNER, W.E., H.A. MOONEY, R.A. GOLDSTEIN, eds.
    1985. Sulfur Dioxide and Vegetation: Ecology,
    Physiology and Policy Issues. Stanford University
    Press, Stanford, California, 593 pp.
11. HECK, W.W., A.S. HEAGLE, D.S. SHRINER. 1986. Effects
    on vegetation: native, crops, forests. In Air
    Pollution. (A.S. Stern, ed.), Vol. 6, Academic
    Press, New York, pp. 247-350.
12. U.S. ENVIRONMENTAL PROTECTION AGENCY. 1980. National
    Air Pollutant Emission Estimates, 1970-1978.
    EPA-450/4-80-002, Office of Air Quality Planning
    and Standards, Research Triangle Park, North
    Carolina.
13. U.S. ENVIRONMENTAL PROTECTION AGENCY. 1978. National
    Air Pollutant Emission Estimates, 1940-1976.

EPA-450/1-78-003, Office of Air Quality Planning
and Standards, Research Triangle Park, North
Carolina.

14. LINTHURST, R.A., A.P. ALTSHULLER, eds. 1984. The
    Acidic Deposition Phenomenon and Its Effects:
    Critical Assessment Document. Vol. 1, U.S.
    Environmental Protection Agency, Washington, D.C.

15. JACOBSON, J.S. 1982. Ozone and the growth and
    productivity of agricultural crops. In M.H.
    Unsworth, D.P. Ormrod, eds., op. cit. Reference 5,
    pp. 293-304.

16. HEATH, R.L. 1980. Initial events in injury to
    plants by air pollutants. Annu. Rev. Plant Physiol.
    31: 395-431.

17. SKARBY, L., G. SELLDEN. 1984. The effects of ozone
    on crops and forests. Ambio 13: 68-72.

18. HEAGLE, A.S., R.B. PHILBECK, H.H. ROGERS, M.B.
    LETCHWORTH. 1979. Dispensing and monitoring
    ozone in open-top field chambers for plant effect
    studies. Phytopathology 69: 15-20.

19. HECK, W.W., O.C. TAYLOR, R. ADAMS, G. BINGHAM, J.
    MILLER, E. PRESTON, L. WEINSTEIN. 1982. Assess-
    ment of crop loss from ozone. J. Air Pollut.
    Control Assoc. 32: 353-362.

20. ADAMS, R.M., S.A. HAMILTON, B.A. McCAROL. 1984. The
    economic effects of ozone on agriculture.
    EPA-600/3-84-090, Environmental Research Laboratory,
    Corvallis, Oregon.

21. REINERT, R.A. 1975. Monitoring, detecting, and
    effects of air pollutants on horticultural crops,
    sensitivity of genera and species. Hort. Sci.
    10: 495-500.

22. ENDRESS, A.G., C. GRUNWALD. 1985. Impact of chronic
    ozone on soybean growth and biomass partitioning.
    Agri. Ecosyst. Environ. 13: 9-23.

23. BLUM, U., E. MROZEK, JR., E. JOHNSON. 1983.
    Investigation of ozone (O_3) effects on [14]C
    distribution in ladino clover. Environ. Exp. Bot.
    23: 369-378.

24. OKANO, K., O. ITO, G. TAKEBA, A. SHIMIZU, T. TOTSUKA.
    1984. Alteration of ^{13}C-assimilate partitioning in
    plants of Phaseolus vulgaris exposed to ozone.
    New Phytol. 97: 155-163.

25. McLAUGHLIN, S.B., R.K. McCONATHY. 1983. Effects of
    SO_2 and O_3 on allocation of ^{14}C-labeled photo-

synthate in Phaseolus vulgaris. Plant Physiol.
73: 630-635.

26. TINGEY, D.T., R.G. WILHOUR, C. STANDLEY. 1976. The
effect of chronic ozone exposures on the metabolite
content of ponderosa pine seedlings. For. Sci.
22: 2334-2341.

27. THOMAS, M.D., G.R. HILL, JR. 1935. Absorption of SO_2
by alfalfa and its relation to leaf injury. Plant
Physiol. 10: 291-307.

28. KATZ, M., G.E. LEDDINGHAM. 1979. Effect of Sulphur
Dioxide on Vegetation. National Research Council
Bulletin 815, Ottawa, Canada, pp. 262-287.

29. BRISLEY, H.R., W.W. JONES. 1950. Sulphur dioxide
fumination of wheat with special reference to its
effect on yield. Plant Physiol. 25: 666-681.

30. ROBERTS, T.M. 1984. Long-term effects of sulfur
dioxide on crops: an analysis of dose-response
relations. Philos. Trans. R. Soc. Lond. Ser. B.
305: 299-316.

31. GUDERIAN, R., H. STRATMANN. 1962. Field experiments
to determine the effects of SO_2 on vegetation.
Part I; Survey of method and evaluation of results.
In Forschungsberichte des Landes Nordrhein -
Westfalen Freilandversuch zur Ermittlung von
Schwefeldioxydwirkungen auf die Vegetation. Part
No. 1118, 102 pp.

32. SPRUGEL, D.G., J.E. MILLER, R.N. MULLER, H.J. SMITH,
P.B. XERIKOS. 1980. Sulfur dioxide effects on
yield and seed quality in field-grown soybeans.
Phytopathology 70: 1129-1133.

33. MILLER, J.E., D.G. SPRUGEL, R.N. MULLER, H.J. SMITH,
P.B. XERIKOS. 1980. Open-air fumigation system
for investigating sulphur dioxide effects on crops.
Phytopathology 70: 1124-1128.

34. HEAGLE, A.S., D.E. BODY, G.E. NEELY. 1974. Injury
and yield responses of soybean to chronic doses of
ozone and sulfur dioxide in the field. Phytopathology 64: 132-136.

35. HEAGLE, A.S., W.W. HECK, J.O. RAWLINGS, R.B. PHILBECK.
1983. Effects of chronic doses of ozone and
sulfur dioxide on injury and yield of soybeans in
open-top chambers. Crop Sci. 23: 1184-1191.

36. MILLER, J.E., H.J. SMITH, W. PREPEJCHAL. 1981.
Evidence for extreme resistance of field corn to
intermittent sulfur dioxide stress. Report No.

ANL-81-85, Part III, Argonne National Laboratory, Argonne, Illinois, pp. 27-29.

37. HECK, W.W., J.A. DUNNING. 1978. Response of oats to sulfur dioxide; interactions of growth temperature with exposure temperature of humidity. J. Air Pollut. Control Assoc. 28: 241-246.

38. FLAGLER, R.B., V.B. YOUNGER. 1982. Ozone and sulfur dioxide effects on tall fescue: I. Growth and yield response. J. Environ. Qual. 11: 386-389.

39. NOYES, R.D. 1980. The comparative effects of sulfur dioxide on photosynthesis and translocation in bean. Physiol. Plant Pathol. 16: 73-79.

40. TEH, K.H., C.A. SWANSON. 1982. Sulfur dioxide inhibition of translocation in bean plants. Plant Physiol. 69: 88-92.

41. JONES, T., T.A. MANSFIELD. 1982. Studies on dry matter partitioning and distribution of ^{14}C labelled assimilates in plants of Phelum pratense exposed to SO_2 pollution. Environ. Pollut. Ser. A 28: 199-208.

42. MILCHUNAS, D.G., W.R. LAURENROTH, J.L. DODD. 1982. The effect of SO_2 on C-14 translocation in Agropyron smithii Rydb. Environ. Exp. Bot. 22: 81-92.

43. REINERT, R.A. 1984. Plant response to air pollutant mixtures. Annu. Rev. Phytopathol. 22: 421-442.

44. ORMROD, D.P. 1982. Air pollution interactions in mixtures. In M.H. Unsworth, D.P. Ormrod, eds., op. cit. Reference 5, pp. 307-331.

45. MENSER, H.A., H. HEGGESTAD. 1966. Ozone and sulfur dioxide synergy injury to tobacco plants. Science 153: 424-425.

46. TINGEY, D.T., R.A. REINERT. 1975. The effect of ozone and sulphur dioxide singly and in combination on plant growth. Environ. Pollut. 9: 117-125.

47. KRESS, L.W., J.E. MILLER, H.J. SMITH. 1986. Impact of ozone and sulphur dioxide on soybean yield. Environ. Pollut. Ser. A 41: 105-123.

48. TINGEY, D.T., W.W. HECK, R.A. REINERT. 1971. Effect of low concentrations of ozone and sulfur dioxide on foliage, growth, and yield of radish. J. Am. Hortic. Sci. 96: 369-371.

49. CONSTANTINIDOU, H.A., T.T. KOZLOWSKI. 1979. Effects of sulfur dioxide and ozone on Ulmus americana seedlings. I. Visible injury and growth. Can. J. Bot. 57: 170-175.

50. HEGGESTAD, H.E., J.H. BENNETT. 1981. Photochemical oxidants potentiate yield losses in snap beans attributable to sulfur dioxide. Science 213: 1008-1010.

51. HOFSTRA, G., D.P. ORMROD. 1977. Ozone and sulphur dioxide interaction in white bean and soybean. Can. J. Plant Sci. 57: 1193-1198.

52. COYNE, P.I., G.E. BINGHAM. 1981. Comparative ozone dose response of gas exchange in a ponderosa pine stand exposed to long-term fumigations. J. Air Pollut. Control Assoc. 31: 38-41.

53. COYNE, P.I., G.E. BINGHAM. 1982. Variation in photosynthesis and stomatal conductance in an ozone-stressed ponderosa pine stand: light response. For. Sci. 28: 257-273.

54. YANG, Y.S., J.M. SKELLY, B.I. CHEVONE, J.B. BIRCH. 1983. Effects of long-term ozone exposure on photosynthesis and dark respiration of eastern white pine. Environ. Sci. Technol. 17: 372-373.

55. REICH, P.B., R.G. AMUNDSON. 1985. Ambient levels of ozone reduce net photosynthesis in tree and crop species. Science 230: 566-570.

56. BLACK, V.J., D.P. ORMROD. M.H. UNSWORTH. 1982. Effects of low concentration of ozone, singly, and in combination with sulphur dioxide on net photo-synthesis rates of Vicia faba L. J. Expt. Bot. 33: 1302-1311.

57. LE SUEUR-BRYMER, N.M., D.P. ORMROD. 1984. Carbon dioxide exchange rates of fruiting soybean plants exposed to ozone and sulfur dioxide singly or in combination. Can. J. Plant Sci. 64: 69-75.

58. CHEVONE, B.I., Y.S. YANG. 1985. CO_2 exchange rates and stomatal diffusive resistance in soybean exposed to O_3 and SO_2. Can. J. Plant Sci. 65: 267-274.

59. PELL, E.J., E. BRENNAN. 1973. Changes in respiration, photosynthesis, adenosing, 5'-triphosphate, and total adenylate content of ozonated pinto bean foliage as they relate to symptom expression. Plant Physiol. 51: 378-381.

60. REICH, P.B. 1983. Effects of low concentrations of O_3 on net photosynthesis, dark respiration and chlorophyll contents in aging hybrid polar leaves. Plant Physiol. 73: 291-296.

61. REICH, P.B., A.W. SCHOETTLE, R.M. RABA, R.G. AMUNDSON. 1986. Response of soybean to low concentrations

of ozone: I. Reductions in leaf and whole plant
net photosynthesis and leaf chlorophyll content.
J. Environ. Qual. 15: 31-36.

62. COYNE, P.I., G.E. BINGHAM. 1978. Photosynthesis and
stomatal light responses in snap beans exposed to
hydrogen sulfide and ozone. J. Air Pollut. Control
Assoc. 28: 1119-1123.

63. TODD, G.W. 1958. Effect of ozone and ozonated 1-
hexene on respiration and photosynthesis of leaves.
Plant Physiol. 416-420.

64. MacDOWALL, H., R.A. LUDWIG. Some effects of ozone
on tobacco leaf metabolism. Phytopathology
(Abstr.) 52: 740.

65. MacDOWALL, F.D.H. 1965. Stages of ozone damage to
respiration of tobacco leaves. Can. J. Bot. 43:
419-427.

66. TODD, G.W., B. PROPST. Changes in transpiration
and photosynthetic rates of various leaves during
treatments with ozonated hexene or ozone gas.
Physiol. Plant. 16: 57-65.

67. DUGGER, W.M., I.P. TING. 1970. Air pollution
oxidants - their effects on metabolic processes
in plants. Annu. Rev. Plant Physiol. 21: 215-234.

68. BARNES, R.L. 1972. Effects of chronic exposure to
ozone on photosynthesis and respiration of pines.
Environ. Pollut. 3: 133-138.

69. HOFSTRA, G., A. ALI, R.T. WUKASCH, R.A. FLETCHER.
1981. The rapid inhibition of root respiration
after exposure of bean (Phaseolus vulgaris L.)
plants to ozone. Atmos. Environ. 15: 483-487.

70. ANDERSON, W.C., O.C. TAYLOR. 1973. Ozone induced
carbon dioxide evolution in tobacco callus cultures.
Physiol. Plant. 28: 419-423.

71. PELL, E.J., W.C. WEISSBERGER. 1976. Histopathological
characterization of ozone injury to soybean foliage.
Phytopathology 66: 856-861.

72. LEE, T.T. 1968. Effect of ozone on swelling of
tobacco mitochondria. Plant Physiol. 43: 133-139.

73. TOMLINSON, H., S. RICH. 1968. The ozone resistance
of leaves as related to their sulfhydryl and
adenosine triphosphate content. Phytopathology 58:
808-810.

74. LEE, T.T. 1967. Inhibition of oxidative phosphorylation and respiration by ozone in tobacco mitochondria. Plant Physiol. 42: 691-696.

75. ITO, O., F. MITSUMORI, T. TOTSUKA. 1984. Effects of NO_2 and O_3 alone or in combination on kidney bean plants. III. Photosynthetic CO_2 assimilation observed by ^{13}C nuclear magnetic resonance. Studies on Effects of Air Pollutant Mixtures on Plants. Part 2. The National Institute for Environmental Studies, No. 66, Japan, pp. 27-36.

76. JOHNSON, E.L. 1984. Effect of ozone on photosynthetic pathways in white clover (Trifolium repens cv) Tillman. Ph.D. Thesis, North Carolina State University, Raleigh, North Carolina.

77. BAILEY, P.S. 1958. The reactions of ozone with organic compounds. Chem. Rev. 58: 926-1110.

78. MUDD, J.B., F. LEH, T.T. McMANUS. 1974. Reaction of ozone with nicotinamide and its derivatives. Arch. Biochem. Biophys. 161: 408-419.

79. MUDD, J.B., R. LEAVITT, A. ONGUN, T.T. McMANUS. 1969. Reaction of ozone with amino acids and proteins. Atmos. Environ. 3: 669-682.

80. HEATH, R.L., P.E. FREDERICK. 1979. Ozone alteration of membrane permeability in Chlorella. Plant Physiol. 64: 455-459.

81. SUTTON, R., I.P. TING. 1977. Evidence for the repair of ozone-induced membrane injury. Amer. J. Bot. 64: 404-411.

82. COULSON, C., R.L. HEATH. 1974. Inhibition of the photosynthetic capacity of isolated chloroplasts by ozone. Plant Physiol. 53: 32-38.

83. NOBEL, P.S., C.T. WANG. 1973. Ozone increases the permeability of isolated pea chloroplasts. Arch. Biochem. Biophys. 157: 388-394.

84. ORDIN, L., M.A. HALL. 1967. Studies on cellulose synthesis by a cell-free oat coleoptile enzyme system: inactivation by airborne oxidants. Plant Physiol. 42: 205-212.

85. ORDIN, L., M.A. HALL, J.I. KINDINGER. 1969. Oxidant-induced inhibition of enzymes involved in cell wall polysaccharide synthesis. Arch. Environ. Health 18: 623-626.

86. DASS, H.C., G.M. WEAVER. 1972. Enzymatic changes in intact leaves of Phaseolus vulgaris following ozone fumigation. Atmos. Environ. 6: 759-763.

87. HANSON, G.P., W.S. STEWART. 1970. Photochemical oxidants: effect on starch hydrolysis in leaves. Science 168: 1223-1224.

88. TINGEY, D.T., R.C. FITES, C. WICKLIFF. 1975. Activity changes in selected enzymes from soybean leaves following ozone exposure. Physiol. Plant. 33: 316-320.

89. CURTIS, C.R., R.K. HOWELL, D.F. KREMER. 1976. Soybean peroxidases from ozone injury. Environ. Pollut. 11: 189-194.

90. MUDD, J.B., T.T. McMANUS, A. ONGUN, T.E. McCULLOGH. 1971. Inhibition of glycolipid biosynthesis in chloroplasts by ozone and sulfhydryl reagents. Plant Physiol. 48: 335-339.

91. TOMLINSON, H., S. RICH. 1973. Anti-senescent compounds reduce injury and steroid changes in ozonated leaves and their chloroplasts. Phytopathology 63: 903-906.

92. GRUNWALD, C., A.G. ENDRESS. 1985. Foliar sterols in soybeans exposed to chronic levels of ozone. Plant Physiol. 77: 245-247.

93. MUDD, J.B. 1982. Effects of oxidants on metabolic functions. In M.H. Unsworth, D.P. Ormrod, eds., op. cit. Reference 5, pp. 189-203.

94. ZIEGLER, I. 1975. The effect of SO$_2$ pollution on plant metabolism. Residue Rev. 56: 79-105.

95. HÄLLGREN, J.-E. 1978. In Sulfur in the Environment. (J.O. Nriagu, ed.), John Wiley and Sons, New York, pp. 163-209.

96. BLACK, V.J. 1982. Effects of sulphur dioxide on physiological processes in plants. In M.H. Unsworth, D.P. Ormrod, eds., op. cit. Reference 5, pp. 67-91.

97. BLACK, V.J., M.H. UNSWORTH. 1979. Effects of low concentrations of sulfur dioxide on net photosynthesis and dark respiration of Vicia faba. J. Exp. Bot. 30: 473-483.

98. WINNER, W.E., H.A. MOONEY. 1980. Ecology of SO$_2$ resistance: II. Photosynthate changes of shrubs in relation to SO$_2$ absorption and stomatal behavior. Oecologia 44: 296-302.

99. BARTON, J.R., S.B. McLAUGHLIN, R.K. McCONATHY. 1980. The effects of SO$_2$ on components of leaf resistance to gas exchange. Environ. Pollut. 21: 255-265.

100. CARLSON, R.W. 1979. Reduction in the photosynthetic rate of Acer, Quercus and Fraxinus species caused by sulphur dioxide and ozone. Environ. Pollut. 18: 159-170.

101. FURUKAWA, A., T. TOTSUKA. 1979. Effects of NO_2, SO_2 and ozone alone and in combinations on net photosynthesis in sunflower. Environ. Control Biol. 17: 161-166.

102. ORMROD, D.P., V.J. BLACK, M.H. UNSWORTH. 1981. Depression of net photosynthesis in Vicia faba L. exposed to sulphur dioxide and ozone. Nature 291: 585-586.

103. THOMAS, M.D., G.K. HILL. 1937. Relation of sulphur dioxide in atmosphere to photosynthesis and respiration in alfalfa. Plant Physiol. 12: 309-383.

104. KATZ, M. 1949. Sulfur dioxide in the atmosphere and its relation to plant life. Ind. Eng. Chem. 41: 2450-2465.

105. SIJ, J.W., C.A. SWANSON. 1974. Short-term kinetic studies on the inhibition of photosynthesis by sulfur dioxide. J. Environ. Qual. 3: 103-107.

106. FURUKAWA, A., T. NATORI, T. TOTSUKA. 1980. The effects of SO_2 on net photosynthesis in sunflower leaf. In Studies on the Effects of Air Pollutants on Plants and Mechanisms of Phytotoxicity. Research Report from the National Institute for Environmental Studies, Vol. 11, Yatabe, Japan, pp. 1-8.

107. TANIYAMA, T., H. ARIKADO, H. IWATA, K. SAWANKA. 1972. Studies on the mechanisms of effects of toxic gases on crop plants. On photosynthesis and dark respiration of rice plant fumigated with SO_2 for long period. Proc. Crop Sci. Soc. Jap. 41: 120-125.

108. VOGL, M., S. BÖRTITIZ. 1965. Physiologische und biochemische Beitrage zur Rauschadenforschung. Flora 155: 347-352.

109. MALHOTRA, S.S. 1976. Effects of sulphur dioxide on biochemical activity and ultrastructural organization of pine needle chloroplasts. New Phytol. 76: 239-245.

110. BALLANTYNE, D.J. 1973. Sulphite inhibition of ATP formation in plant mitochondria. Phytochemistry 12: 1207-1209.

111. MALHOTRA, S.S., D. HOCKING. 1976. Biochemical and cytological effects of sulphur dioxide on plant metabolism. New Phytol. 16: 227-237.

112. HARVEY, G.W., A.H. LEGGE. 1979. The effect of sulfur dioxide upon metabolic level of adenosine triphosphate. Can. J. Bot. 57: 759-764.

113. ZELITCH, I. 1971. In Photosynthesis, Photorespiration and Plant Productivity. Academic Press, New York, pp. 130-212.

114. KOZIOL, M.J., C.F. JORDON. 1978. Changes in carbohydrate levels in red kidney bean (Phaseolus vulgaris L.) exposed to sulfur dioxide. J. Exp. Bot. 29: 1037-1043.

115. KOZIOL, M.J., D.W. COWLING. 1978. Growth of ryegrass (Lolium perenne L.) exposed to SO_2. II. Changes in the distribution of photoassimilated ^{14}C. J. Exp. Bot. 29: 1431-1439.

116. LIBERA, W., I. ZIEGLER, H. ZIEGLER. 1974. The action of sulfite on the HCO_3-fixation and the fixation pattern of isolated chloroplasts and leaf tissue slices. Z. Pflanzenphysiol. 74: 420-433.

117. SPEDDING, D.J., W.J. THOMAS. 1973. Effect of sulphur dioxide on the metabolism of glycolic acid by barley (Hordeum vulgare) leaves. Aust. J. Biol. Sci. 26: 281-286.

118. SOLDATINI, G.F., I. ZIEGLER. 1979. Induction of glycolate oxidase by $SO2$ in Nicotiana tabacum. Phytochemistry 18: 21-22.

119. GUDERIAN, R., H. van HAUT. 1970. Detection of SO_2 effects upon plants. Staub-Reinhalt. Luft 30: 22-35.

120. PUCKETT, K.J., D.H.S. RICHARDSON, W.P. FLORA, E. NIEBOER. 1974. Photosynthetic ^{14}C fixation by the lichen Umbilicaria muhlenbergii (Ach.) tuck. following short exposures to aqueous sulphur dioxide. New Phytol. 73: 1183-1192.

121. NIEBOER, E., D.H.S. RICHARDSON, K.J. PUCKETT, F.O. TOMASSINI. 1976. Phytotoxicity of sulfur dioxides in relation to measurable response in lichens. In Effects of Air Pollutants on Plants. (T.A. Mansfield, ed.), Cambridge University Press, Cambridge, England, pp. 61-86.

122. ALSCHER-HERMAN, R. 1982. The effect of sulphite on light activation of chloroplast fructose 1,6-bisphosphatase in two cultivars of soybean. Environ. Pollut. Ser. A 27: 83-96.

123. LeBLANC, F., D.N. RAO. 1974. A review of the literature on bryophytes with respect to air pollution. Bull. Soc. Bot. Fr., Colloq. Bryol. 121: 237-255.

124. MALHOTRA, S.S. 1977. Effects of aqueous sulphur dioxide on chlorophyll destruction in Pinus contorta. New Phytol. 78: 101-109.

125. TANAKA, H., I. TAKANASHI, M. YATAZAWA. 1972. Experimental studies on sulfur dioxide injuries in higher plants. I. Formation of glyoxylate-bisulfite in plant leaves exposed to sulfur dioxide. Water Air Soil Pollut. 1: 205-211.

126. ZELITCH, I. 1957. β-Hydroxysulfonates as inhibitors of the enzymatic oxidation of glycolic and latric acids. J. Biol. Chem. 224: 251-260.

127. PAUL, J.S., J.A. BASSHAM. 1978. Effects of sulfate on metabolism in isolated mesophyll cells from Papaver somniferum. Plant Physiol. 62: 210-214.

128. OSMOND, C.B., P.N. AVADHANI. 1970. Inhibition of the β-carboxylation pathway of CO_2 fixation by bisulfite compounds. Plant Physiol. 45: 228-230.

129. ZIEGLER, I. 1973. Effect of sulphite on phospho-enolpyruvate carboxylase and malate formation in extracts of Zea mays. Phytochemistry 12: 1027-1030.

130. ZIEGLER, I. 1972. The effect of SO_3--on the activity of ribulose-1,5-diphosphate carboxylase in isolated spinach chloroplasts. Planta 103: 155-163.

131. GEZEILUS, K., J.-E. HALLGREN. 1980. Effect of SO_3^{-2} on the activity of ribulose bisphosphate carboxylase from seedlings of Pinus silvestris. Physiol. Plant. 49: 354-358.

132. PARRY, M.A.J., S. GUTTERIDGE. 1984. The effect of SO_3^{-2} and SO_4^{-2} ions on the reactions of ribulose bisphosphate carboxylase. J. Exp. Bot. 35: 157-168.

133. OLEKSYN, J. 1984. Effects of SO_2, HF and NO_2 on net photosynthetic and dark respiration rates of scots pine needles of various ages. Photosynthetica 18: 259-262.

134. MULLER, R.N., J.E. MILLER, D.G. SPRUGEL. 1979. Photosynthetic response of field-grown soybeans to fumigations with sulphur dioxide. J. Appl. Ecol. 16: 567-576.

135. CARLSON, R.W. 1983. The effect of SO_2 on photo-synthesis and leaf resistance at varying concen-trations of CO_2. Environ. Pollut. Ser. A 39: 309-322.

136. KATASE, M., T. USHIJIMA, T. TAZAKI. 1983. The relationship between absorption of sulfur dioxide

(SO₂) and inhibition of photosynthesis in several plants. Bot. Mag. 96: 1-13.

137. TAYLOR, JR., G.E., W.J. SELVIDGE. 1984. Phytotoxicity in bush bean of five sulfur-containing gases released from advanced fossil energy technologies. J. Environ. Qual. 13: 224-230.

138. TEMPLE, P.J., O.C. TAYLOR, L.F. BENOIT. 1985. Effects of ozone on yield of two field-grown barley cultivars. Environ. Pollut. Ser. A 39: 217-225.

139. KOHUT, R., J.A. LAURENCE. 1983. Yield response of red kidney bean, Phaseolus vulgaris, to incremental ozone concentrations in the field. Environ. Pollut. Ser. A 32: 233.

140. KRESS, L.W., J.E. MILLER. 1985. Impact of ozone on field-corn yield. Can. J. Bot. 63: 2408-2415.

141. TEMPLE, P.J., O.C. TAYLOR, L.F. BENOIT. 1985. Cotton yield responses to ozone as mediated by soil moisture and evapotranspiration. J. Environ. Qual. 14: 55-60.

142. HEAGLE, A.S., M.B. LETCHWORTH, C.A. MITCHELL. 1983. Injury and yield responses of peanuts to chronic doses of ozone in open-top field chambers. Phytopathology 73: 551-555.

143. KRESS, L.W., J.E. MILLER. 1985. Impact of ozone on grain sorghum yield. Water Air Soil Pollut. 25: 377-390.

144. KRESS, L.W., J.E. MILLER. 1983. Impact of ozone on soybean yield. J. Environ. Qual. 12: 276-281.

145. HEAGLE, A.S., V.M. LESSER, J.O. RAWLINGS, W.W. HECK, R.B. PHILBECK. 1986. Response of soybeans to chronic doses of ozone applied as constant or proportional additions to ambient air. Phytopathology 76: 51-56.

146. HEGGESTAD, H.E., T.J. GISH, J.H. BENNETT, L.W. DOUGLASS. 1982. Influence of soil moisture stress on the response of soybeans to O₃ x SO₂ doses. In National Crop Loss Assessment Network (NCLAN) 1982 Annual Report. (W.W. Heck, O.C. Taylor, R.M. Adams, G.E. Bingham, J.E. Miller, E.M. Preston, L.H. Weinstein, eds.), Corvallis Environmental Research Laboratory, Office of Research and Development, U.S. Environmental Protection Agency, Corvallis, Oregon, pp. 111-143.

147. KOHUT, R.J., R.G. AMUNDSON. 1981. Impact of ozone on soybean yield, growth and net photosynthesis. In National Crop Loss Assessment Network (NCLAN)

1981 Annual Report. (W.W. Heck, O.C. Taylor, R.M. Adams, G.E. Bingham, J.E. Miller, E.M. Preston, L.H. Weinstein, eds.), Corvallis Environmental Research Laboratory, Office of Research and Development, U.S. Environmental Protection Agency, Corvallis, Oregon, pp. 51-76.

148. KRESS, L.W., J.E. MILLER, H.J. SMITH. 1985. Impact of ozone on winter wheat yield. Environ. Exp. Bot. 25: 211-228.

149. HECK, W.W., W.W. CURE, J.O. RAWLINGS, L.J. ZARAGOZA, A.S. HEAGLE, H.E. HEGGESTAD, R.J. KOHUT, L.W. KRESS, P.J. TEMPLE. 1984. Assessing impacts of ozone on agricultural crops: II. Crop yield functions and alternative exposure statistics. J. Air Pollut. Control Assoc. 34: 810-817.

150. TINGEY, D.T., R.A. REINERT, C. WICKLIFF, W.W. HECK. 1973. Chronic ozone or sulfur dioxide exposures, or both, affect the early vegetative growth of soybean. Can. J. Plant Sci. 53: 875-879.

151. SHIMIZU, H., S. MOTOHASHI, H. IWAKI, A. FURUKAWA, T. TOTSUKA. 1981. Effects of chronic exposures to ozone on the growth of sunflower plants. Environ. Control Biol. 19: 137-147.

152. FOSTER, K.W., H. TIMM, C.K. LABANAUSKAS, R.J. OSHIMA. 1983. Effects of ozone and sulfur dioxide on tuber yield and quality of potatoes. J. Environ. Qual. 12: 75-79.

153. BENNETT, J.P., R.J. OSHIMA. 1976. Carrot injury and yield response to ozone. J. Am. Soc. Hort. Sci. 101: 638-639.

154. ITO, O., K. OKANO, M. KUROIWA, T. TOTSUKA. 1985. Effects of NO_2 and O_3 alone or in combination on kidney bean plants (Phaseolus vulgaris L.): growth, partitioning of assimilates and root activities. J. Exp. Bot. 36: 652-662.

155. BENNETT, J.P., V.C. RUNECKLES. 1977. Effects of low levels of ozone on growth of crimson and annual ryegrass. Crop Sci. 17: 443-445.

156. BENNETT, J.P., R.J. OSHIMA, L.F. LIPPERT. 1979. Effects of ozone on injury and dry matter partitioning in pepper plants. Environ. Exp. Bot. 19: 33-39.

157. OSHIMA, R.J., P.K. BRAEGELMANN, R.B. FLAGER, R.R. TESO. 1979. The effects of ozone on the growth, yield and partitioning of dry matter in cotton. J. Environ. Qual. 8: 474-479.

158. REICH, P.B., J.P. LASSOIE. 1985. Influence of low concentrations of ozone on growth, biomass partitioning and leaf senescence in young hybrid poplar plants. Environ. Pollut. Ser. A 39: 39-51.

159. OSHIMA, R.J., J.P. BENNETT, P.K. BRAEGELMANN. 1978. Effect of ozone on growth and assimilate partitioning in parsley. J. Am. Soc. Hort. Sci. 103: 348-350.

160. LINZON, S.N. 1971. Economic effects of sulfur dioxide on forest growth. J. Air Pollut. Control Assoc. 21: 81-86.

161. NEELY, G.E., D.T. TINGEY, R.G. WILHOUR. 1977. Effects of ozone and sulfur dioxide singly and in combination on yield quality, and N-fixation of alfalfa. In International Conference on Phytochemical Oxidant Pollution and Its Control. (B. Dimitriades, ed.), EPA-600/3-77-001b, EPA, Research Triangle Park, North Carolina, pp. 663-673.

162. BELL, J.N.B., W.S. CLOUGH. 1973. Depression of yield in ryegrass exposed to sulphur dioxide. Nature 241: 47-49.

163. DOCHINGER, L.S., K.F. JENSEN. 1975. Effects of chronic and acute exposure to sulphur dioxide on the growth of hybrid poplar cuttings. Environ. Pollut. 9: 219-229.

164. COWLING, D.W., D.R. LOCKYER. 1978. The effect of SO_2 on Lolium perenne grown at different levels of sulphur and nitrogen nutrition. J. Exp. Bot. 29: 257-265.

165. BELL, J.N.B., A.J. RUTTER, J. RELTON. 1979. Studies on the effects of low levels of sulphur dioxide on the growth of Lolium perenne L. New Phytol. 83: 627-643.

166. LAURENCE, J.A. 1979. Response of maize and wheat to sulfur dioxide. Plant Dis. Rep. 63: 468-471.

167. ASHENDEN, T.W., I.A.D. WILLIAMS. 1980. Growth reduction in Lolium multiforum Lam. and Phleum preatense L. as a result of SO_2 and NO_2 pollution. Environ. Pollut. Ser. A 21: 131-139.

168. REINERT, R.A., D.E. WEBER. 1980. Ozone and sulfur dioxide-induced changes in soybean growth. Phytopathology 70: 914-916.

169. NORBY, R.J., T.T. KOZLOWSKI. 1981. Response of SO_2-fumigated Pinus resinosa seedlings to

postfumigation temperatures. Can. J. Bot. 59:
470-475.

170. DUGGER, JR., W.M., J. KOUKOL, R.L. PALMER. 1966.
Physiological and biochemical effects of
atmospheric oxidants on plants. J. Air Pollut.
Control Assoc. 16: 467-471.

171. DUGGER, JR., W.M., R.L. PALMER. 1969. Carbohydrate
metabolism in leaves of rough lemon as influenced
by ozone. Proc. First Intl. Citrus Symp., Vol. 2,
pp. 711-715.

172. MILLER, P.R., F.W. COBB, JR., E. ZAVARIN. 1968.
III. Effect of injury upon oleoresin composition,
phloem carbohydrates, and phloem pH. Hilgardia
39: 135-140

173. MILLER, P.R., J.R. PARMETER, JR., B.H. FLICK, C.W.
MARTINEZ. 1969. Ozone dosage response of
ponderosa pine seedlings. J. Air Pollut.
Control Assoc. 19: 435-438.

174. PIPPEN, E.L., A.I. POTTER, V.G. RANDALL, K.G. NG,
I.W. REUTER, III, A.G. MORGAN, JR., R.J. OSHIMA.
1975. Effect of ozone fumigation on crop composition. J. Food Sci. 40: 672-676.

175. TINGEY, D.T., R.C. FITES, C. WICKLIFF. 1973. Ozone
alteration of nitrate reduction in soybean.
Physiol. Plant. 29: 33-38.

176. WILKINSON, T.G., R.L. BARNES. 1973. Effects of
ozone on $^{14}CO_2$ fixation patterns in pine. Can. J.
Bot. 51: 1573-1578.

177. PELL, E.J., N.S. PEARSON. 1984. Ozone-induced
reduction in quantity and quality of two potato
cultivars. Environ. Pollut. Ser. A 35: 345-352.

178. BÖRTITZ, S. 1969. Physiologische und biochemische
Beiträge aur Rauchschadenforschung. Arch.
Forstwes. Bd. 18: 123-131.

179. CONSTANTINIDOU, H.A., T.T. KOZLOWSKI. 1979.
Effects of sulfur dioxide and ozone on Ulmus
americana seedlings. II. Carbohydrates, proteins,
and lipids. Can. J. Bot. 57: 176-184.

180. KOZIOL, M.J., D.W. COWLING. 1980. Growth of ryegrass (Lolium perenne L.) exposed to SO_2. J.
Exp. Bot. 31: 1687-1699.

181. MALHOTRA, S.S., S.K. SARKAR. 1979. Effects of
sulfur dioxide on sugar and free amino acid
content of pine seedlings. Physiol. Plant. 47:
223-228.

182. BLUM, U., G.R. SMITH, R.C. FITES. 1982. Effects of multiple O_3 exposures on carbohydrate and mineral contents of ladino clover. Environ. Expt. Bot. 22: 143-154.

Chapter Four

UPTAKE AND METABOLISM OF PHENOLIC COMPOUNDS BY THE WATER
HYACINTH (Eichhornia crassipes)

DAVID H. O'KEEFFE*, THOMAS E. WIESE*,
SHAUNA R. BRUMMET+ AND TODD W. MILLER+

*Department of Chemistry
The University of Michigan-Flint
Flint, Michigan 48502

+Department of Chemistry
The University of Akron
Akron, Ohio 45235

INTRODUCTION

The water hyacinth (Eichhornia crassipes) (Fig. 1) is
a free-floating, subtropical, freshwater macrophyte. It
is considered to be one of the world's worst aquatic
weeds. This is in no small part due to its phenomenal
growth rate. It has been estimated that 10 adult plants
can multiply to about 650,000 in only eight months.[1]
Large mats of water hyacinth restrict navigation for
transportation and recreation, impede drainage, greatly
hasten eutrophication, and increase water loss by evapotranspiration.

Fig. 1.   The water hyacinth (<u>Eichhornia</u> <u>crassipes</u>) after
Penfound and Earle[1] (reproduced with permission of the
Ecological Society of America).

     The potential to use water hyacinths in wastewater
treatment comes from the same characteristics that make
the plant a nuisance – its vigor and fast growth. When
introduced into sewage lagoons water hyacinths grow well
and reduce the biological oxygen demand, total suspended
solids, and total nitrogen and phosphorus content.[2-5]
This aquatic weed is also very effective at reducing the
nutrient load of the water in which they grow.[6-10] In
fact water hyacinth treatment provides tertiary water
quality at below conventional treatment costs.[11,12]
However, it is the water hyacinth's ability to help in
the treatment of toxic and/or hazardous environmental
pollutants which provides us with the greatest incentive
to carefully examine its phytochemistry.[13,14]

     This plant has already been shown to bioconcentrate
many metal ions from water. Examples include lead,
cadmium, mercury, copper, zinc, and chromium.[14-20] Much
work continues on the mechanisms of uptake,[21-24] and
translocation.[15] Storage of such metal ions in the plant
and the potential for their recovery also needs investi-

PHENOL

HYDROXYBENZENE

R (-Cl,-CH$_3$,-NO$_2$)

MONOSUBSTITUTED PHENOLS

CATECHOL

1,2-DIHYDROXYBENZENE

HO

RESORCINOL

1,3-DIHYDROXYBENZENE

HYDROQUINONE

1,4-DIHYDROXYBENZENE

PYROGALLOL

1,2,3-TRIHYDROXYBENZENE

PHLOROGLUCINOL

1,3,5-TRIHYDROXYBENZENE

HYDROXYQUINOL

1,2,4-TRIHYDROXYBENZENE

Fig. 2. Chemical structures of phenolic compounds.

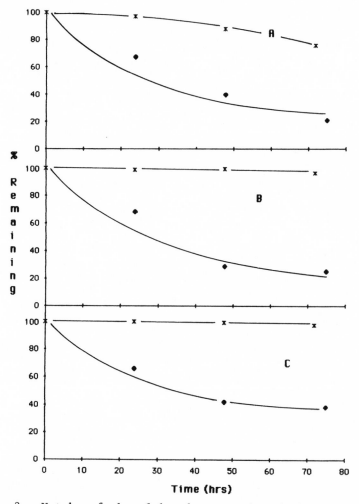

Fig. 3. Uptake of phenol by the water hyacinth. Initial
concentrations were 50 mg/L (A), 100 mg/L (B) and 400 mg/L
(C) phenol containing 2-3 x 10^{-3} μCi U-^{14}C phenol in 50%
Hoagland's solution (400 mL). Percent remaining was
determined by liquid scintillation spectrometry: (⨉)
without plant, (◆) with plants.

gation.   Uptake of organic compounds has received
considerably less attention.   The removal from solution
of a few herbicides,[25],[26] mirex and toxaphene[27],[28] and
mevinphos[29] has been briefly examined.

Water hyacinths have also been reported to take up
phenol.[30]   In addition, a simple water hyacinth-based
unit for the continuous removal of phenol from solution
has been designed.[31]   This chemical and a myriad of
synthetic phenolic compounds are common environmental
pollutants.[32]   They are known to arise from, among
other sources, the coking of coal, gas works and oil
refineries, chemical and pesticide plants as well as dye
manufacturing plants.[33]   When present in the aquatic
ecosystem they represent a toxic threat to most plant and
animal life.   Thus we have initiated studies that will
allow us to determine the extent to which the water
hyacinth is capable of taking up and metabolizing (detoxifying?) this class of environmental pollutants.   The
structures of the phenolic compounds used in this study
are shown in Figure 2.   The results presented here
suggest that this aquatic weed, a nuisance plant, may
become very important in the control of organic compounds
in the environment.

EXPOSURE TO AND TOXICITY OF PHENOLIC COMPOUNDS

Phenol

The uptake of $U-^{14}C$-phenol at 50, 100 and 400 mg/L
in Hoagland's solution by the water hyacinth is shown in
Figure 3.   These plots are similar to those reported by
Wolverton and McKown.[30]   The rate and extent of phenol
removal from solution were observed to vary as a function
of plant size, temperature, and time of year; however the
results shown in Figure 3 are fairly typical.   Uptake data
obtained by the radiotracer technique and by high performance liquid chromatography (reverse-phase C-18 or cyano
column; mobile phase of methanol;water, 50:50 or 60:40 v/v
containing 1% acetic acid; 1 mL/min flow rate; 280 nm
detection wavelength) are summarized in Table 1.   The
results are in excellent agreement and indicate that
either method may be used to monitor the water hyacinth's
ability to take up a variety of phenolic compounds from
aqueous solution.

Table 1. Uptake of Phenol. Comparison of scintillation
spectrometry and HPLC methods for determination of phenol.

Time(hrs)	LSS				HPLC			
	Plants	Dev.	Control	Dev.	Plants	Dev.	Control	Dev.
0	50	3	50	3	50	2	50	2
5	44	3	50	2	40	4	48	0.5
24	28	6	42	2	30	3	47	0.5
48	17	5	45	1	12	4	48	0.3
72	12	6	42	2	9	4	45	2

Time(hrs)	LSS				HPLC			
	Plants	Dev.	Control	Dev.	Plants	Dev.	Control	Dev.
0	200	11	200	11	200	14	200	14
24	142	13	199	10	151	2	200	3
48	104	15	194	9	99	2	197	3
72	83	32	195	10	75	6	198	5

LSS: Liquid scintillation spectrometry. Phenol determined as $U-^{14}C$ phenol.

HPLC: High performance liquid chromatography. Phenol determined directly
by the area percent quantitation method.

All data as mg/L ± standard deviation.

    Phenol loss from control solutions was observed in
all uptake experiments (Fig. 3, Table 1), reminiscent
of the Wolverton and McKown experiments.[30] These investigators analyzed for bacteria in water samples taken from
plant controls, phenol controls, water controls and
plant-phenol containers. An insignificant variation in
bacterial counts was shown. In the present studies
bacterial counts as bacteria per millilitre (BPM) were
made from yeast mannitol agar plates streaked at times
0, 24, 48, and 72 hours in autoclaved solutions of uptake
experiments (50 and 200 mg/L phenol). In autoclaved
control containers bacterial growth occurred beginning at
48 hours, due to contamination during sampling. Experimental containers first showed growth at 24 hours (50 mg/L).
Although bacterial growth is documented to occur (Table 2)
in control containers, the decrease in phenol was only 10%
at 50 mg/L and 1% at 200 mg/L (Table 1). This loss of
phenol was probably due to bacterial metabolism. Bacterial
growth in experiments was greater than in controls, but
it was not enough to account for the phenol decrease

Table 2. Bacterial counts in sterilized containers and Hoagland's solutions containing phenol.[a]

TIME (hours)

	0	24	48	72
Control 50 mg/L	0[b]	0	0	3.5×10^3
Expt. 50 mg/L	0	4.8×10^3	4.9×10^5	5.0×10^5
Control 200 mg/L	0	0	2.5×10^4	5.0×10^4
Expt. 200 mg/L	0	0	3.0×10^4	1.0×10^5
Plant Control		1.0×10^6	5.0×10^6	9.0×10^6

[a]50% Hoagland's solution (pH 7)

[b]Bacteria per millilitre (BPM) determined by dilution and plating on yeast mannitol agar

observed in these containers. Therefore the decrease in phenol in experimental containers can be accounted for almost entirely by the water hyacinth. Phenol also appears to be toxic to bacteria as indicated by the results in Table 2. At all exposure times both experimental and phenol control containers have much lower bacterial loads as compared to a plant control.

The acutely toxic phenol concentration for plants weighing one to three grams (dry weight) was determined to be about 200 mg/L. Plants exposed to this or higher concentrations did not survive beyond four days after exposure to phenol. However, the plants decreased the amount of phenol in solution by 60-80% within 72 hours and did not return it to solution when they died. Concentrations below this level are toxic to plants but usually not lethal to them. The outward signs of phenol toxicity are withering and drying of the leaves, without chlorosis, similar to that occurring when the solution is allowed to evaporate. The lower and outermost leaves are the first affected. The roots also show marked response to phenol. The normally white to pale purplish roots become dark burnt brown in color. They become quite brittle after several days and do not appear to be alive. Plants exposed to 50 mg/L phenol or greater showed these effects within 24 hours. The solution also turns brown - the higher the initial phenol concentration, the darker the color (24 hours). Although the solution does not appear turbid, most of the color at 24 hours is due to suspended solid material since it is retained on a 0.45 μm nylon filter when the solution is filtered.

Plants subsequently exposed to identical or higher levels of phenol continue to remove it from solution (see Tolerance Development). New root growth becomes evident after several days. Such roots are usually several times thicker than the original ones, do not initially possess many root hairs and are not brown in color. Also, considerably less leaf damage is noted during such repeat exposures.

Dihydroxybenzenes

Representative plots illustrating time versus amount remaining when water hyacinths are exposed to 50 mg/L of catechol, hydroquinone and resorcinol are shown in Figure 4. Tables 3-5 list the time versus amount remaining in solution for plants exposed to a range of concentrations (50-200 mg/L). At an exposure level of 50 mg/L, catechol is initially cleared from solution faster than phenol with twice as much catechol cleared in 6 hours (76%, Fig. 4) compared to phenol (37% under identical conditions). Catechol uptake leveled off more quickly than phenol, though, and the average uptake at 48 hours--while still greater than 90% of the initial catechol present--was slightly less than the phenol uptake at 48 hours. In 68 hours, the exposed plants were able to remove almost all of the catechol in the 100 mg/L solution but not the 200 mg/L solution. At 67.5 hours, the uptake of catechol from the 200 mg/L solutions had essentially stopped at a catechol concentration of about 70 mg/L (Table 3) and the exposed plants were in poor health.

Overall, hydroquinone is taken up from solution by the water hyacinth. Figure 4 shows the average values obtained from five studies of water hyacinth exposure to 50 mg/L solutions. Although control values drop a relatively large amount, the solution values indicate water hyacinth uptake of hydroquinone similar to that seen for phenol and catechol. At about 48 hours, over 98% of the

Table 3. Uptake of catechol.

TIME(hrs)	PLANTS n	(mg/L)	DEV	CONTROL n	(mg/L)	DEV	PLANTS n	(mg/L)	DEV	CONTROL n	(mg/L)	DEV
0	2	100.0	6.9		100.0		8	201.4	6.1		201.4	
6												
12	3	62.1	13.2	1	95.0		7	129.2	14.2	5	202.0	0.7
26	3	53.5	18.0	1	83.8		7	105.0	12.8	5	186.4	4.0
46	3	36.6	3.4	1	95.0		11	81.6	20.4	9	198.3	2.0
68	1	1.2		1	95.4		5	70.2	2.7	5	196.0	1.0

Table 4. Uptake of hydroquinone.

TIME(hrs)	PLANTS n	(mg/L)	DEV	CONTROL n	(mg/L)	DEV	PLANTS n	(mg/L)	DEV	CONTROL n	(mg/L)	DEV
0	14	100.0	12.0		100.0		20	199.2	8.7		199.2	
4							4	193.9	1.0		-	
6												
12	11	91.0	2.6	9	95.9	1.1	15	194.5	6.4	13	194.0	2.8
26	11	81.5	6.8	9	93.0	1.4	15	142.4	29.5	13	172.2	11.7
49	11	23.3	7.4	9	81.0	6.0	19	56.8	41.3	17	170.7	8.4
69	1	7.5	*	1	103.6	*	5	21.6	0.3	5	128.2	0.4

*Single value.

‾No sample taken.

initial hydroquinone had been cleared from solution.   It
is also cleared readily from solutions of 100 and 200 mg/L
(Table 4).

Data collected from three studies of water hyacinth
exposure to solutions of resorcinol at 50 and 200 mg/L are
shown in Table 5. No uptake of resorcinol was observed
and the 200 mg/L data show a definite increase in resorcinol
concentration with time. This increased concentration is
real and due to the decrease in solution volume during the
exposure study. The relatively rapid evapotranspiration
rate of the unaffected plants in the resorcinol exposure

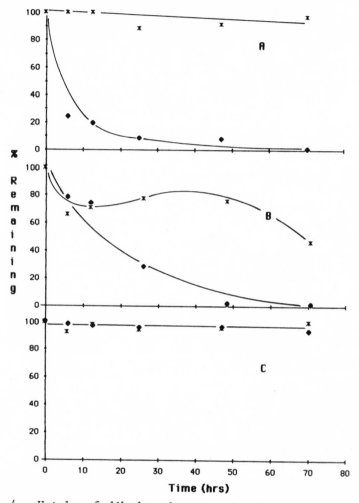

Fig. 4. Uptake of dihydroxybenzenes by the water hyacinth.
Initial concentrations of catechol (A), hydroquinone (B)
and resorcinol (C) were 50 mg/L in Hoagland's solution.
Percent remaining was determined by HPLC: (X) without
plant, (◆) with plants.

Table 5. Exposure to resorcinol.

TIME(hrs)	PLANTS			CONTROL			PLANTS			CONTROL		
	n	(mg/L)	DEV	n	(mg/L)	DEV	n	(mg/L)	DEV	n	(mg/L)	DEV
0	10	50.3	2.6		50.3		8	199.3	5.6		199.3	
4							4	197.8	0.6		*	
6	8	49.2	1.3	8	46.4	4.4						
12	11	48.9	2.9	9	49.1	3.3	7	191.8	4.8	5	195.1	1.9
26	11	48.1	1.5	9	47.3	1.0	7	198.7	5.1	5	195.0	2.2
46	11	48.1	3.9	9	48.0	1.5	11	213.4	4.0	9	198.5	4.2
69	9	47.1	2.6	9	50.3	3.6	5	228.0	1.9	5	209.4	1.0

*No sample taken

decreased the solution volume while the mass of resorcinol present remained constant, resulting in an increased concentration of resorcinol.

At initial exposures to solutions of hydroquinone greater than 200 mg/L, the plants did not survive; catechol was acutely toxic at less than 200 mg/L. The toxic effects observed were some yellowing of the floats, stems, and leaves (chlorosis) accompanied by wilting and eventual plant death. The plant roots and rhizome turned dark brown or black, became mushy and began to decay when exposed to these acutely toxic solutions. The toxic effects were observed to be concentration dependent - quite similar to the toxicity of phenol. That catechol is more toxic and hydroquinone less toxic than phenol was also shown by examining their concentration dependence. After 24 hours of exposure to 25 mg/L catechol, plants suffered considerable leaf and float damage along with root darkening. Exposure of plants to the same amount of hydroquinone resulted in minimal darkening of the roots and hardly any toxic effects to the leaves and floats. Plants exposed to higher levels of these two dihydroxy-benzenes suffered increasingly greater damage to roots, leaves and floats. Examination of the solutions from these exposures after 24 hours also yielded interesting results. As in the case of phenol, a concentration dependent color development was observed. Solutions from plants exposed to catechol were lavender-purple in color while those resulting from hydroquinone exposure were yellow-brown. Filtration through 0.45 μm nylon filters yielded solutions having considerably less intense color and solid material on the filters.

Exposure of water hyacinths to solutions of resorcinol in the 25–200 mg/L range did not result in any apparent toxic effects, consistent with the absence of uptake (Table 5). Exposure to solutions of resorcinol did kill the water hyacinths when initial concentrations were well above 1000 mg/L although the mechanism of plant death seemed to be simple dehydration. The stems and leaves of plants exposed to resorcinol dried to a crispy texture with no color changes, suggesting that the high concentration of resorcinol prevented the normal uptake of water. To rank the compounds discussed so far, catechol is the most toxic, showing acute toxicity at initial exposures less than 200 mg/L, phenol is next at 200 mg/L, and hydroquinone is only toxic at exposure levels greater than 200 mg/L. Resorcinol is essentially non-toxic to water hyacinths until an exposure level of 2000 mg/L is reached.

Trihydroxybenzenes

Based upon the results from the exposure of water hyacinths to the three dihydroxybenzenes, predictions for uptake and toxicity of the trihydroxybenzenes can be made. Phloroglucinol, the all-meta (1,3,5) isomer, should not be taken up and should not affect water hyacinths exposed to it. Both 1,2,3- and 1,2,4- trihydroxybenzenes (pyrogallol and hydroxyquinol) should be taken up from solution by the water hyacinth and should be toxic to the exposed plants.

Water hyacinths were exposed to deionized water solutions of each of the trihydroxybenzenes in one study lasting for four days. Because of this and the need for more development of the chromatography of these compounds, the results here should be considered preliminary until confirmed by further work. The uptake predicted for pyrogallol and hydroxyquinol was demonstrated in exposure experiments at 50 and 100 mg/L solutions of those compounds; the data for 50 mg/L are shown in Figure 5. Water hyacinths exposed to a 100 mg/L solution of phloroglucinol responded as predicted by not taking up the compound and by continuing to transpire water at a rapid rate. The predicted pattern of toxicity of the trihydroxybenzenes was confirmed,

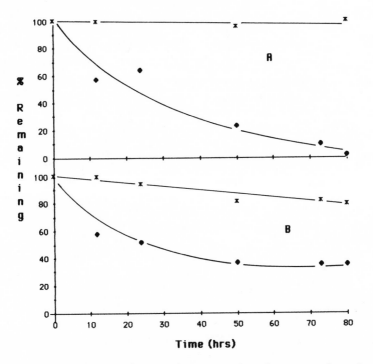

Fig. 5. Uptake of trihydroxybenzenes by the water hyacinth. Initial concentrations of pyrogallol (A) and hydroxyquinol (B) were 50 mg/L in Hoagland's solution. Percent remaining was determined by HPLC: (X) without plants, (◆) with plants.

but the differences among the compounds were more pronounced. Pyrogallol (1,2,3-trihydroxybenzene) was clearly the most toxic; only one of the twelve plants exposed survived after the study (the surviving plant had been exposed to the 25 mg/L solution). Conversely, all but one of the water hyacinths exposed to 1,2,4-trihydroxybenzene (hydroxyquinol) survived; and the dead plant came from the solution containing 100 mg/L. Also as predicted, exposure to 1,3,5-trihydroxybenzene at these levels had no deleterious effects on the water hyacinths.

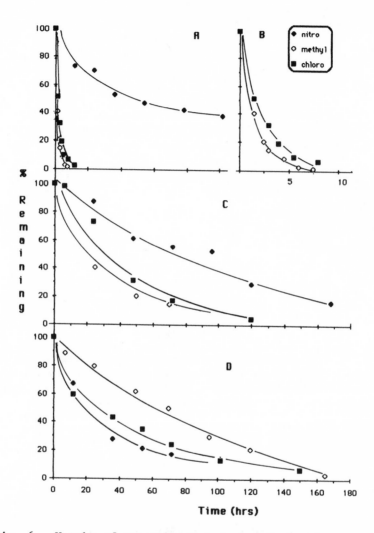

Fig. 6. Uptake of monosubstituted phenols by the water
hyacinth. Initial concentrations of each isomer of
chlorophenol (■), methylphenol (◇) and nitrophenol (◆)
were 50 mg/L in deionized water. Percent remaining was
determined by HPLC. The para isomers (A and B) are shown
on two time scales; meta isomers (C) and ortho isomers
(D) on a single scale.

Monosubstituted Phenols

Data regarding the uptake of the isomeric forms (o, m, p) of chloro-, methyl- and nitrophenols by water hyacinth plants exposed to 50 mg/L concentrations of each is shown in Figure 6. All of these compounds are at least partially removed from deionized water or Hoagland's solution in the presence of plants. The p-chloro- and p-methylphenols are removed from solution very rapidly (Fig. 6A and B). Less than 1% of each remains in solution after a 15-hour exposure. The meta and ortho isomers of these two phenols are taken up much more slowly. The time required to see a 50% decrease in concentration of their meta isomers is 20-25 hours and 35-40 hours for their ortho isomers. The rate of uptake of the three isomers of the chloro- and methylphenols decreases in the order p > m > o.

Each of the three nitrophenols is also taken up by the water hyacinth. The meta and ortho isomers appear to require 75-80 and 20-25 hours, respectively, for removal of 50% (Fig. 6C and D). Preliminary kinetic analyses of phenolic compound uptake (Table 6) suggest that the times are 60.3 (meta) and 47.6 hours (ortho). These times are somewhat longer (a factor of three for the meta isomers) than those for the chloro- and methylphenols. The uptake of p-nitrophenol is very slow, the slowest of all phenolic compounds examined thus far. The apparent time for 50% uptake is 40-45 hours (Fig. 6A) while 112 hours is predicted from the kinetic analysis (Table 6). Thus the water hyacinth deals with exposure to p-nitrophenol quite differently than with exposure to the para isomers of chloro- and methylphenols.

The most obvious initial sign of toxicity to the water hyacinths exposed to the chloro-, methyl- and nitrophenols is the on-set of lower leaf wilting (dehydration). As noted with the other phenolic compounds the rate at which this leaf damage occurred followed quite closely the rate of uptake of these compounds. The 50 mg/L exposure level proved to be lethal to almost every plant. However, detailed experimental evaluation of chemical structure and positional isomerism along with exposure levels to both acute and sub-acute toxicity to plants has not been completed.

Table 6. Kinetic data from phenolic compound uptake experiments.

Phenolic Compound	Correlation Coefficient	Rate Constant (k) hr^{-1}	Time for 50% decrease ( ), hrs	Time for 99% decrease, hrs
phenol	0.960	0.962	7.20	50.6
	0.971	0.641	10.8	75.4
	0.999	0.580	12.0	65.6
o-methylphenol	0.944	0.020	34.6	250
o-chlorophenol	0.997	0.0170	40.7	256
o-nitrophenol	0.956	0.0146	47.6	277
m-methylphenol	0.979	0.0310	22.5	169
m-chlorophenol	0.998	0.0286	24.2	169
m-nitrophenol	0.972	0.0115	60.3	417
p-methylphenol	0.993	0.608	1.1	7.5
p-chlorophenol	0.985	0.307	2.3	13.8
p-nitrophenol	0.955	0.00619	112	692

One interesting observation related to positional isomerism is seen after 28 hours exposure; solutions of p-chloro- and p-methylphenol and the roots of the plants being exposed developed color. Solutions of p-nitrophenol exposed to plants always retained some of their initial color as described below. Exposure solutions from the meta and ortho isomers of all three of these monosubstituted phenols stayed clear and colorless throughout, and all plant roots retained their normal appearance. The solution of p-chlorophenol became light to medium brown in color, as did the roots of the plant. The p-methylphenol solution became medium orange in color with the plant's roots becoming light orange-brown. This behavior is similar to that involving phenol, catechol and hydroquinone exposure solutions and plant roots, and, as observed in those cases, solid material was retained on 0.45 μm nylon filters after passing through aliquots (20 ml) of p-chloro- and p-methyl-phenol solutions. At the 28-hour exposure time the p-nitrophenol solution retained its very distinctive yellow-green color indicative of the phenolate ion. The plant's roots also seemed to have been partially bleached; that is, they appeared lighter than when the exposure was initiated.

KINETICS OF PHENOLIC COMPOUND UPTAKE

A preliminary examination of the uptake rate data obtained from exposing water hyacinth plants to a number of different phenolic compounds reveals consistent first order kinetics (Table 6). The calculated correlation coefficients range from a high of 0.998 (m-chlorophenol) to a low of 0.944 (o-methylphenol). Since the very large numbers of plants of same age and size required for sophisticated rate studies is an impracticality, most experiments have been done in triplicate (at the same time) to yield average data. Nevertheless, it has been consistently demonstrated that experiments performed at different times during the year can yield quite inconsistent rate data. Amid this inconsistency (range?), however, is the typical first order uptake behavior as shown in Figure 7 for the isomeric forms of chlorophenol. The results listed in Table 6 for the uptake of the parent phenol compound represent the fairly typical range from various times during the year. The uptake experiments (average of two or three) from which the data were obtained for the kinetic analyses of the chloro-, methyl- and nitrophenols were all performed within a five-week period.

The trends in the kinetic data with regard to effects of positional isomerism and substituent identity were fairly consistent. The p-methyphenol ($t_{1/2}$ = 1.1 hours) and p-chlorophenol isomers ($t_{1/2}$ = 2.3 hours) were always the most rapidly taken up by the water hyacinths, even more rapidly than phenol itself ($t_{1/2} \sim$ 10 hours). p-Nitrophenol was always the most slowly cleared from solution ($t_{1/2}$ = 112 hours), with its meta isomer being the next slowest ($t_{1/2}$ = 60.3 hours). Positional isomerism yielded interesting results. The uptake rates decreased in the order para > meta > ortho for the chloro- and methylphenols while the order for the nitrophenols was just the opposite (o>m> p), although the differences among $t_{1/2}$ values were not as large (Table 6). Yet another trend noted for each positional isomer of the three monosubstituted phenols is that the rate at which they were cleared from solution decreased in the substituent order $-CH_3$ > $-Cl$ > $-NO_2$.

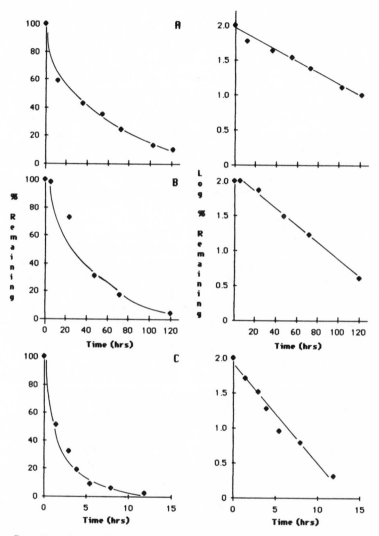

Fig. 7. Uptake and first order kinetic plots for the three chlorophenols: ortho isomer (A), meta isomer (B) and para isomer (C).

TOLERANCE DEVELOPMENT TO PHENOL AND CATECHOL

The ability of the water hyacinth to develop a tolerance to phenol levels both below and above the acutely toxic concentration was also demonstrated. Plants were first placed in Hoagland's solution containing 50 or 100 mg/L phenol. The phenol concentration was then raised in steps of 50 or 100 mg/L every 24, 48, or 72 hours. In one experiment the initial concentration was 100 mg/L and it was increased every 72 hours. The maximum concentration at which the plants could survive was found to be about 400 mg/L. When this level was reached in this series of experiments it was maintained by replacing the 400 mg/L phenol in Hoagland's solution every two or three days. In this manner, over a period of three weeks, a single water hyacinth plant weighing one to two grams (dry weight) took up 0.40 to 0.45 grams of phenol. Plants treated in this way still find phenol to be toxic; however the damage is considerably reduced from that observed in initial exposures to solutions of > 200 mg/L phenol.

Plants placed in easily tolerated (< 100 mg/L) phenol concentrations were observed to produce new roots as previously noted. As the phenol concentration was continuously increased, the production of new roots also continued. The new roots were not damaged by the higher (> 200 mg/L) phenol levels when compared with roots placed directly into high levels.

The results presented here represent a doubling of the concentration of phenol that can be tolerated by the plant simply by gradually increasing the level of phenol. In other words, the organism was induced to tolerate normally toxic levels of phenol. The second important fact presented in these studies is that the plants lived in and continuously took up concentrations of phenol toxic to most plants for a period of three weeks. At this time the experiment was terminated. The plants, however, could have continued living in the phenol solutions for an indeterminate period. The rate of uptake of the high levels (after conditioning) appeared to be nearly linear and reached an average rate of about 2.5 mg/L phenol removed per hour.

Plants exposed to fresh 50 mg/L catechol in Hoagland's solution suffered no ill effects and continued each day to

clear all the catechol from solution each day throughout
a twelve-day experiment. In all, four water hyacinths in
two liters of solution cleared 1.0 g of catechol in twelve
days.

That the water hyacinths which found 200 mg/L solutions
of catechol acutely toxic can be acclimated to more than
twice that level has also been demonstrated. In another
twelve-day experiment water hyacinths were conditioned to
survive in a solution of 450 mg/L and to continue to clear
the catechol from that solution. Plants were initially
exposed to a 25 mg/L level for one day and then were transferred daily in steps of 25-50 mg/L to increasing catechol
levels. The rate of uptake also increased in a manner
similar to that noted in the phenol experiment. At day
twelve of the study, all four water hyacinths had taken up
about 2 g of catechol, showed no evidence of wilting or
chlorosis, continued to grow new roots and leaves, and
appeared healthy.

IDENTIFICATION OF A PHENOL METABOLITE

A cell-free enzyme supernatant (250 mL) from water
hyacinth root tissue (46 g) was prepared for use in examining the metabolic fate of phenol. An experiment was
designed to determine whether or not phenol is initially
oxidized to catechol. Thus the idea was to attempt to
trap ^{14}C-labeled catechol formed from U-^{14}C-phenol. The
cell-free supernatant was thoroughly oxygenated. Phenol
(100 mg/L) was added along with 25 µCi of U-^{14}C-phenol.
Since it was expected that any catechol formed would be
very rapidly converted to other metabolites, an excess of
this compound (1000 mg/L) was added to the supernatant.
Any ^{14}C-catechol produced would be greatly diluted in the
excess catechol present and thus trapped. The presence of
radioactivity in the catechol isolated from the supernatant
would be evidence that ^{14}C-catechol was indeed formed from
the labeled phenol. Samples of aerated supernatant were
incubated at 25°C for two and ten minutes. Hydrochloric
acid (2N) was added to stop enzyme activity and the
samples were extracted with ethyl ether (three volumes) to
recover phenolic compounds. Separation of the remaining
phenol, catechol and other phenolic compounds was done by
HPLC using a semi-preparative reverse-phase C-18 column.
The chromatograms are shown in Figure 8.

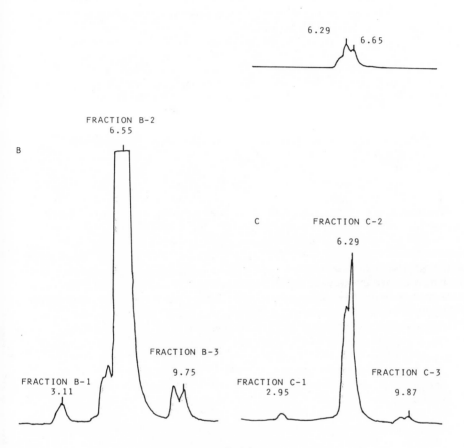

TIME (min)

Fig. 8. HPLC chromatograms of ether extracts from a cell-free suspension (root tissue) of water hyacinth as a function of time of exposure to phenol. Traces as a function of time (minutes) of exposure to phenol: (A) control - 0 min, (B) 2 min, (C) 10 min. Separation was performed on a Hewlett-Packard RP-C_{18} semi-preparative column (7 μm).

The chromatogram of a sample extracted from the acidified cell-free supernatant prior to addition of phenol,
U-^{14}C-phenol and catechol is shown in Figure 8A. Figures
8B and 8C are the chromatograms from samples at two and
ten minutes. Fractions B-3 and C-3 are phenol which was
added to the original solution. Fractions B-2 and C-2
are catechol. Verification of the identities of these
fractions consisted of chromatographing each of them, as
well as phenol and catechol standards, on the RP C-18
analytical column. The retention times of fraction B-2
and catechol agree, well within the 3% limits for catechol
(Fig.9) as do their UV spectra, each taken with the HPLC's
internal scanning spectrophotometer. Both have maxima
at 276 nm as well as a shoulder at 214 nm. Identical UV
spectra were recorded on a Beckman Model 25 UV-VIS
scanning spectrophotometer.

The entire volumes of each phenolic compound containing
ether extracts from the two and ten-minute incubations were
subjected to chromatography on the semi-preparative column.
The combined eluates consisting of the B fractions from
the two-minute and C fractions from the ten-minute incubations, respectively, were analyzed using liquid scintillation spectroscopy. The radioactivity in fractions B-2
and C-2 combined (i.e., the catechol fractions) was
determined to be approximately 10% of the total label
recovered. The combined fractions B-3 and C-3, representing
phenol, accounted for about 85% of the total. The
remainder of the radioactivity (5%) was present in the
combination of fractions B-1 and C-1, which presumably
represents unidentified, even more polar metabolite(s).
Although these results must be considered as only
preliminary since the experiment has been performed just
once, it seems reasonable to conclude that at least a
position of the original phenol was metabolized by the
plant to catechol.

CONCLUSIONS

It is known that the polyphenol oxidase enzymes in
plants can metabolize monophenols (catechol oxidase/
cresolase activity), o-diphenols (catechol oxidase/
catecholase activity) plus p-diphenols and m-diphenols
(laccase).[34,35] In fact it has been suggested that this
system be used to remove phenol from wastewaters.[36] The

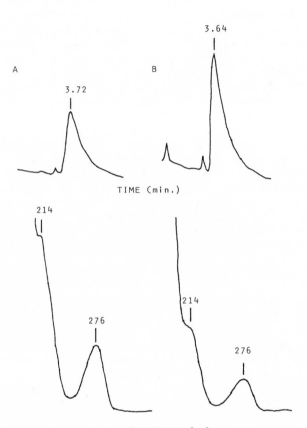

Fig. 9. Verification of fraction B-2 as catechol (see text and Fig. 8). Analytical HPLC chromatograms (RP, C-18 column) of catechol (A) and fraction B-2 (B) are shown at the top. UV spectra (recorded by the chromatograph's internal scanning spectrophotometer) of catechol (A) and fraction B-2 (B) are shown at the bottom.

discoloration of the water hyacinth roots and of the
solutions in exposure studies where phenol, catechol,
hydroquinone, p-chloro and p-methylphenols were examined
for uptake suggests that damage to root tissue probably
occurred. The extent of this damage has not as yet been
assessed. However, it would seem as though polyphenoloxidases present in root tissue act upon these particular
exogenous phenolics. The observation that this group of
compounds is comprised of both mono- (including the chloro
and methyl p-substituted phenols) and p-diphenols suggests
that the root tissue has more than one type of copper
oxidase present. Related metabolic processes have been
reported. Seidel[37-41] has shown that the bulrush
(Scirpus) removes phenolic compounds from solution and
has preliminary evidence indicating that phenol injected
into the rhizome or the basal meristem is metabolized.
Stom and co-workers have briefly discussed the toxicity
of polyphenols to a few aquatic plants.[42] They have also
suggested that the elimination of exogenous phenols by
selected hydrophytes is consistent with the action of
polyphenoloxidases.[43-46]

It is of some interest that the phenolics giving rise
to color in the root tissues and in solution are by far
the most toxic of the compounds investigated thus far.
They are also the most rapidly taken up from solution,
perhaps as the result of being converted to solid material
in solution and presumably in the root tissue. Yet the
loss of these phenolics from solution cannot be explained
this easily. The development of color in the root tissue
and in solution is concentration dependent. The compounds
are removed from solution even at levels where no color is
observed. Repeated exposures of a plant to concentrations
high enough to cause color development initially yield
less or even no color. The metabolic fate(s) of the
phenolics under these conditions is unknown. Furthermore,
these additional exposures (even at higher concentrations)
are much less toxic to the plant. The phytochemistry
involved here is currently being investigated.

It must be noted that the apparent inability of the
water hyacinth to remove resorcinol (1,3-dihydroxybenzene)
from solution is not readily explained. Perhaps a laccase-
type oxidase is not present. However, hydroquinone, a
p-diphenol and a substrate for such an enzyme, was rapidly
removed from solution. The nitrophenols could be acting
as uncouplers as well as inhibitors of polyphenoloxidases

but that fails to explain their differential rate of uptake by the water hyacinth. Finally, the fate(s) of the o- and m-chlorophenols as well as the o- and m-methylphenols is (are) not yet known. Much additional work remains in order to elucidate the phytochemical effects of phenolic compounds on the water hyacinth.

ACKNOWLEDGMENTS

The authors would like to thank Rachel Jarvis for performing some of the experiments, Barbara O'Keeffe and Julie Lynch for help in preparing the figures and Doritta McDaniel for typing the manuscript. The authors also acknowledge financial support in the form of Faculty Research and Development Grants from The University of Michigan-Flint and The University of Akron. Additional support was obtained from The State of Michigan Research Excellence and Economic Development Fund (administered by The Project for Urban and Regional Affairs, The University of Michigan-Flint).

REFERENCES

1. PENFOUND, W.T., T.T. EARLE. 1948. The biology of the water hyacinth. Ecol. Monogr. 18: 447-472.
2. GOSSETT, D.R., W.E. NORRIS, JR. 1972. Relationship between nutrient availability and content of nitrogen and phosphorus in tissues of the aquatic macrophyte Eichhornia crassipes (Mart.) Solms. Hydrobiologia 38: 15-28.
3. McDONALD, R.C., B.C. WOLVERTON. 1980. Comparative study of wastewater lagoon with and without water hyacinth. Econ. Bot. 34: 101-110.
4. DeBUSK, T.A., L.D. WILLIAMS, J.H. RYTHER. 1983. Removal of nitrogen and phosphorus from waste- water in a water hyacinth-based treatment system. J. Environ. Qual. 12: 257-262.
5. BOYD, C.E. 1970. Vascular aquatic plants for mineral nutrient removal from polluted waters. Econ. Bot. 24: 95-103.
6. SHEFFIELD, C.W. 1967. Water hyacinths for nutrient removal. Hyacinth Control J. 6: 27-30.
7. ROGERS, H.H., D.E. DAVIS. 1972. Nutrient removal by water hyacinth. Weed Sci. 20: 423-428.

8. CORNWALL, D.A., J. ZOLTEK, JR., C.D. PATRINELY, T.
    deS. FURMAN, J.I. KIM. 1977. Nutrient removal
    by water hyacinths. J. Water Pollut. Control.
    Fed. 49: 57–65.
9. WOLVERTON, B.C., R.C. McDONALD. 1979. Upgrading
    facultative wastewater lagoons with vascular
    aquatic plants. J. Water Pollut. Control Fed.
    51: 305–313.
10. REDDY, K.R., J.C. TUCKER. 1983. Productivity and
    nutrient uptake of water hyacinth, Eichhornia
    crassipes. I. Effect of nitrogen source. Econ.
    Bot. 37: 237–247.
11. SIMMONDS, M.A. 1979. Tertiary treatment with
    aquatic macrophytes. Prog. Water Technol. 11:
    507–518.
12. TRIDECH, S., A.J. ENGLANDE, JR., M.J. HERBERT, R.F.
    WILKINSON. 1981. Tertiary wastewater treatment
    by the application of vascular plants. In
    Chemistry of Water Reuse. (W.J. Cooper, ed.),
    Vol. 2, Ann Arbor Science Publishers, pp. 521–539.
13. WOLVERTON, B.C., R.M. BARLOW, R.C. McDONALD. 1976.
    Application of vascular aquatic plants for
    pollution removal, energy and food production in
    a biological system. In Biological Control of
    Water Pollution. (J. Tourbier, R.W. Pierson, Jr.,
    eds.), University of Pennsylvania Press, pp.
    141–149.
14. WOLVERTON, B.C., R.C. McDONALD. 1979. The water
    hyacinth: from prolific pest to potential
    provider. Ambio 8: 2–9.
15. CHIGBO, F.E., R.W. SMITH, F.L. SHORE. 1982. Uptake
    of arsenic, cadmium, lead and mercury from
    polluted waters by the water hyacinth Eichhornia
    crassipes. Environ. Pollut. Ser. A 27: 31–36.
16. COOLEY, T.N., D.F. MARTIN. 1977. Factors affecting
    the distribution of trace elements in aquatic
    plants. J. Inorg. Nucl. Chem. 39: 1893–1896.
17. JANA, S., M.A. CHOUDURI. 1984. Synergistic effects
    of heavy metal pollutants and senescence in
    submerged aquatic plants. Water Air Soil
    Pollut. 21: 351–357.
18. KAY, S.H., W.T. HALLER, L.A. GARRARD. 1984. Effects
    of heavy metals on water hyacinths (Eichhornia
    crassipes (Mart.) Solms). Aquat. Toxicol.
    (Amsterdam) 5: 117–128.

19. MURAMOTO, S., Y. OKI. 1983. Removal of some heavy metals from polluted water by water hyacinth (Eichhornia crassipes). Bull. Environ. Control Biol. 39: 170-177.

20. LOW, K.S., C.K. LEE. 1981. Cooper, zinc, nickel and chromium uptake by "kangkong air" (Ipomea aquatica Forsk). Pertanika 4: 16-20.

21. O'KEEFFE, D.H., J.K. HARDY, R.A. RAO. 1984. Cadmium uptake by the water hyacinth: effects of solution factors. Environ. Pollut. Ser. A 34: 133-147.

22. HARDY, J.K., D.H. O'KEEFFE. 1985. Cadmium uptake by the water hyacinth: effects of root mass, solution volume, complexers and other metal ions. Chemosphere 14: 417-426.

23. HARDY, J.K., N.B. RABER. 1985. Zinc uptake by the water hyacinth: effects of solution factors. Chemosphere 14: 1155-1166.

24. HEATON, C., J. FRAME, J.K. HARDY. 1986. Lead uptake by Eichhornia crassipes. Toxicol. Environ. Chem. 11: 125-136.

25. BINGHAM, S.W. 1973. Improving water quality by removal of pesticide pollutants with aquatic plants. Office of Water Resources Research (PB219389), 94 p.

26. BINGHAM, S.W., R.L. SHAVER. 1977. Diphenamid removal from water and metabolism by aquatic plants. Pestic. Biochem. Physiol. 7: 8-15.

27. SMITH, R.W., F.L. SHORE. 1978. Water hyacinths for removal of mirex from water. J. Miss. Acad. Sci. Suppl. 23: 22.

28. SMITH, R.W., W. VAN ZANDT, F.L. SHORE. 1977. Water hyacinths for removal of toxaphene from water. J. Miss. Acad. Sci. Suppl. 22: 20.

29. WOLVERTON, B.C. 1975. Aquatic plants for removal of mevinphos from aquatic environment. NASA Tech. Memo. TM-X-72720.

30. WOLVERTON, B.C., M.M. McKOWN. 1976. Water hyacinths for removal of phenols from polluted waters. Aquat. Bot. 2: 191-201.

31. VAIDYANATHAN, S., K.M. KAVADIA, M.G. RAO, S. BASU, S.P. MAHAJAN. 1983. Removal of phenol using water hyacinth in a continuous unit. Int. J. Environ. Stud. 21: 183-191.

32. DeWALLE, F.B., D.A. KALMAN, R. DILLS, D. NORMAN, E.S.K. CHIAN, M. GIABBAI, M. GHOSAL. 1982. Presence

of phenolic compounds in sewage, effluent and
sludge from municipal sewage treatment plants.
Water Sci. Technol. 14: 143-150.

33. BUIKEMA, A.L., JR., M.J. McGINNISS, J. CAIRNS, JR.
1979. Phenolics in aquatic ecosystems: a
selected review of recent literature. Mar.
Environ. Res. 2: 87-181.

34. MAYER, A.M., E. HAREL. 1979. Polyphenoloxidases
in plants. Phytochemistry 18: 193-215.

35. VAMOS-VIGYAZO, L. 1981. Polyphenol oxidase and
peroxidase in fruits and vegetables. CRC Crit.
Rev. Food Sci. Nutr. 15: 49-127.

36. VEDRALOVA, E., Z. PECHAN, J. DUCHON. 1980. Removing
phenol from waste waters by oxidation to melanin
with mushroom polyphenoloxidase. Collect.
Czech. Chem. Commun. 45: 623-627.

37. SEIDEL, K. 1963. On phenol accumulation and phenol
reduction in water plants. Naturwissenschaften
50: 452-453.

38. SEIDEL, K. 1965. Phenol reduction in water by
Scirpus lacustris L. during a 31-month investigation. Naturwissenschaften 52: 398.

39. SEIDEL, K. 1966. Purification of surface water by
higher plants. Naturwissenschaften 53: 289-297.

40. SEIDEL, K. 1967. Mixotrophy in Scirpus lacustris L.
Naturwissenschaften 54: 176-177.

41. SEIDEL, K. 1976. Macrophytes and water purification.
In Biological Control of Water Pollution. (J.
Tourbier, R.W. Pierson, Jr., eds.), University of
Pennsylvania Press, pp. 141-149.

42. STOM, D.I., R. ROTH. 1981. Some effects of polyphenols on aquatic plants: I. Toxicity of phenols
in aquatic plants. Bull. Environ. Contam.
Toxicol. 27: 332-337.

43. STOM, D.J., S.S. TIMOFEEVA, N.F. KASHINA, L.J.
BIELYKH, S.N. SOUSLOV, V.V. BOUTOROB, M.S.
APARTZIN. 1980. Methods of analyzing quinones
in water and their application in studying the
effects of hydrophytes on phenols. Part 3:
Phenol elimination under the action of aquatic
plants. Acta hydrochim. hydrobiol. 8: 223-230.

44. STOM, D.J., S.S. TIMOFEEVA, N.F. KASINA, L.J.
BIELYKH, S.N. SOUSLOV, V.V. BOUTOROV, M.S.
APARTZIN. 1980. Methods of analyzing quinones
in water and their application in studying the
effects of hydrophytes on phenols. Part 4:

Accumulation of exogenic phenols, localization of endogenic phenols and o-diphenol oxidase in Nitella cells. Acta hydrochim. hydrobiol. 8: 231-240.

45. STOM, D.J., S.S. TIMOFEEVA, N.F. KASHINA, L.J. BIELYKH, S.N. SOUSLOV, V.V. BOUTOROV AND M.S. APARTZIN. 1980. Methods of analyzing quinones in water and their application in studying the effects of hydrophytes on phenols. Part 5: Elimination of carcinogenic amines from solutions under the action of Nitella. Acta hydrochim. hydrobiol. 8: 241-245.

46. STOM, D.J., S.S. TIMOFEEVA, S.N. SOUSLOV. 1981. Some methods of phenol elimination from sewage waters. Part 1: Biodestruction by the vegetable homogenates. Acta hydrochim. hydrobiol. 9: 433-445.

Chapter Five

CONTROL OF TRACE ELEMENT TOXICITY BY PHYTOPLANKTON

JAMES G. SANDERS AND GERHARDT F. RIEDEL

The Academy of Natural Sciences
Benedict Estuarine Research Laboratory
Benedict, Maryland   20612

INTRODUCTION

The transport and impact of toxic substances are matters of major concern to management of coastal zones. Man's love for the coastline and its attendant bays and estuaries ensures that, as population and industrial growth continue, the loading of toxic substances into estuaries and the coastal ocean will continue to increase. It is imperative, therefore, that society address the potential toxic effect of anthropogenic inputs to these areas. However, there are two impediments to successfully addressing these concerns.

First, there are an overwhelming number of potential toxic substances. The large number of toxic inorganic compounds, and the ever-increasing array of organic compounds require that initial assessments of potential toxic substances be made, on the basis of their occurrence, concentration, and whatever is known about their toxicity.[1-3]

Second, it is difficult to assess accurately the potential for toxicity at a community, or worse, ecosystem level. Once toxics that pose potential problems have been identified, there remains the additional problem of the assessment itself. There is increasing appreciation

that laboratory determination of the levels of toxic
substances sufficient to kill a few species in bioassays
under carefully controlled conditions will not provide
adequate understanding of the environmental consequences
after release of the same substances. Organisms exhibit
widely varying sensitivity to different types of toxic
materials. In addition, physical and geochemical processes
within the ecosystem greatly affect the transport,
reactivity, and potential impact that may occur. Often
impact will not be manifested as a simple decrease in
productivity, but rather as sublethal, subtle changes in
the structure of the biological communities that comprise
an estuarine ecosystem. Biological processes occurring
within organisms also can transform the original compound
into a new set of compounds that may have quite different
chemical and biological properties.

     To illustrate the complexity of this problem, we offer
the following example. Marine research has focused
recently on the interaction of algae and inorganic ions,
particularly metals. Research efforts have centered on
the effect of metal species on algal nutrition, growth,
and metabolism. We are beginning to understand the
complexity of interaction between specific metal ions
and cellular biochemistry, interactions which are of
paramount importance to the physiological ecology of
algae in aquatic systems.

     Algal cells are able to discriminate between a wide
variety of chemical compounds, actively transporting and
incorporating the elements necessary for cell growth,
such as N, P, Mn, Fe, and Zn. However, chemically similar
ions may be confused and the "competing" ion taken
up indiscriminately. Competition between chemical
analogues for uptake or attachment sites has been
described for a number of ion pairs, such as cadmium/
nitrate,[4] copper/silicate,[5,6] copper/manganese,[7] arsenate/
phosphate,[8,9] and chromate/sulfate.[10] If the competing
ion is inhibitory to growth, as is the case with the above,
cellular incorporation of this ion will be detrimental to
the success of the algal species. Some species are better
able to discriminate between competing ion pairs. Thus,
the ability to discriminate between competing ions becomes
an important factor in the continuation of a particular
species as a member of the phytoplankton community.

Algae, in turn, play an important role in the geochemistry of many trace ions. Many elements are actively involved in biological processes and are taken up by algal cells, leading to their removal from surface waters (i.e., Cu, Zn, Se). These elements are then incorporated into particulate material, affecting their transport and reactivity. Some elements also undergo considerable restructuring of chemical form after biological uptake. With toxic elements, this change in form can drastically alter biological "availability" and toxicity; thus, this biological role is a very important one.

This last-discussed phenomenon, the control of trace element availability and toxicity by phytoplankton, is not well understood and is the subject of this article. We will focus on two, quite different toxic elements-- arsenic and chromium. Arsenic is extremely toxic, biologically reactive, and present at elevated levels in many estuaries and coastal zones. Additionally, arsenic is present in a number of chemical forms which greatly affect its toxicity, transport, and reactivity. Chromium is also quite toxic and is biologically reactive; however, it has a much different geochemistry from arsenic.

ARSENIC BIOGEOCHEMISTRY AND TOXICITY

Arsenic is ubiquitous in nature, present in water and soils and concentrated in a variety of ores. It is quite toxic to organisms. Emissions of arsenic and subsequent anthropogenic input of arsenic to the oceans has increased greatly during the past 100 years.[11-13] Current predictions indicate that arsenic concentrations in surface waters will continue to increase.

A number of estuaries already receive large quantities of arsenic from man; these include Puget Sound,[14,15] the Tejo Estuary in Portugal,[16] and the Tamar River and other estuaries in southwest England.[17,18] Recent data from the Chesapeake Bay indicate that daily arsenic loadings to this estuary from man's activities exceed 100 kg per day.[19] Continued increases have the potential for altering biological processes in marine systems; however, our basic knowledge of arsenic geochemistry presently is

inadequate to predict with confidence whether such
alterations will have any significance. Although global
arsenic geochemistry is fairly well described,[13] the
movement and transformations of arsenic within aquatic
systems, especially estuaries, is poorly understood.

Arsenic in oxidized, aquatic systems is present
primarily as an inorganic ion, arsenate,[14,20-22] with the
predominant dissolved form being $HAsO_4^-$.[23,24] Reduced
arsenic [arsenite, As(III)] and methylated arsenicals
[methylarsonic (MMA) and dimethylarsinic (DMA) acids] are
also present occasionally.

The production of reduced and methylated species is
mediated by biological processes.[9,20,25] Arsenate is
chemically similar to phosphate and is readily taken up
by phytoplankton,[9] but not retained in large quantities.
Most of the arsenic is released to the surrounding water
in chemically altered form, either reduced or methylated.

The quantity and chemical form of arsenic released,
and resulting concentrations of reduced and methylated
species, vary between ecosystems. The variation in
reduced arsenic is well correlated with primary productivity.[26,27] Reduced and methylated species are more
prevalent during algal blooms, often comprising up to
20% of the total arsenic concentration in coastal areas[22]
and up to 80% in productive estuaries such as the
Chesapeake Bay.[27]

The differences between estuarine and oceanic arsenic
biogeochemistry may result from fundamental differences
in the adaptation of the respective phytoplankton
communities to variation in nutrient concentrations.
Estuarine phytoplankton communities generally are
dominated by rapidly-growing, opportunistic species with
low nutrient affinities. This growth strategy is
successful in estuarine areas where low physical stability
and relatively high nutrient concentrations prevail.
Oceanic communities, on the other hand, are dominated
by species with high affinity for nutrients, adapted for
the more stable, low nutrient concentrations found in
open oceans.[28,29] Species that exhibit high affinity for
nutrients appear to be better able to discriminate between
the necessary nutrient and competing ions than do species
exhibiting lower nutrient affinity. This has been

demonstrated with both natural phytoplankton communities[30] and for individual species.[31] Increased discrimination should effectively exclude arsenate from uptake.[30] This variation in nutrient affinity in general suggests that estuarine communities will take up greater quantities of arsenic which they must then dispose of by reduction, methylation, and release.

The proportion of reduced and methylated arsenic species is of extreme importance to the ecosystem as a whole. Arsenite and methylarsonic species have been shown to be the most toxic arsenic forms.[32-34] Arsenite is unstable under oxidizing conditions and is rapidly reconverted to arsenate.[35-37] However, both MMA and DMA are relatively stable in marine systems, and may persist indefinitely.[38,39] The production, release and persistence of MMA in a coastal ecosystem may be extremely important, not only because it relieves phytoplankton of the inhibitory effects of arsenate but also because it increases the toxic burden to other organisms within the ecosystem. Thus, other biota within the ecosystem may be inhibited by the change in chemical form and resultant increased toxicity.

Arsenate at very low concentrations (<5 µg per liter) significantly affects the growth rate of a number of marine phytoplankton.[40-42] However, the degree of inhibition is species specific (Table 1). Because of this differential sensitivity, chronic inputs of low concentrations of arsenate can alter dominant species and species succession within a natural phytoplankton community.[42-44]

There are two mechanisms by which algal species may succeed and dominate after exposure to arsenic stress; only one mechanism requires arsenic resistance.

1. Successful species may have a higher affinity for phosphorus (see above), and thus a lower incorporation rate of arsenate because of a higher degree of specificity for the uptake of phosphate relative to arsenate.[31] Although slower growing, these algal forms would not be susceptible to increased concentrations of arsenate and would be able to dominate over more rapidly growing cells

Table 1. Sensitivity of phytoplankton clones to arsenate.

Taxonomic Class, Species	Clone	EC_{50} (μM)	Reference
Dinophyceae			
Amphidinium carterae	AMPHI	0.13	42
Peridinium trochoidium	PT	1.2	36
Chrysophyceae			
Isochrysis galbana	ISO	0.03	42
Bacillariophyceae			
Rhizosolenia fragilissima	RFRAG	0.03	44
Skeletonema costatum	SKEL	0.07	41
Cerataulina pelagica	–	0.08	43
Chaetoceros pseudocrinitum	CH1	0.27	42
Thalassiosira pseudonana	W	>0.35	This study
Thalassiosira pseudonana	SWAN1	>0.35	This study
Thalassiosira pseudonana	3H	>1.3	42
Cylindrotheca closterium	NC	>1.3	42
Chlorophyceae			
Tetraselmis contracta	TETRA	>1.3	42

The arsenate concentrations listed are those necessary to cause a 50% reduction in growth rate, and include background concentrations of arsenate (~ 0.013 μM).

that were incorporating larger quantities of arsenate.

2. Some species of algae may be able to efficiently process incorporated arsenic, allowing them to survive and perhaps dominate even though their

arsenic incorporation rates are comparable to previous dominants. Some algal species are very efficient at transforming arsenate and releasing it in altered form; others are not.[9,20,45,46] Indeed, unusual chemical forms of arsenic in natural systems have been correlated with single species of algae.[19,27] Because these transformations require energy, cell growth may be inhibited somewhat, but the species can continue to exist within the community at lowered cell densities. The occurrence of large quantities of reduced and methylated arsenic in productive estuaries is likely due to this second mechanism.

The transformation of arsenic by phytoplankton is quite variable. Centric diatoms seem to produce arsenite and DMA predominantly in culture; some species produce small amounts of MMA.[9,20,27,45] Coccoliths grown under similar conditions also produce MMA.[45] Until recently, however, only arsenite and DMA had been detected as by-products of natural phytoplankton blooms; MMA had not been detected in significant concentrations.[20,22,39,47] These observations were consistent with recent studies implicating methylarsonic acid as an intermediate step in reductive methylation of arsenate to DMA by algae.[46] Therefore, MMA would not be excreted by algae. However, we have been sampling the Chesapeake Bay system since 1980 and have often found quite high concentrations of MMA, up to 400 ng per liter, which is correlated with the presence of one algal species.[19] Although DMA, on average, is the predominant reduced form present, MMA is present in most samples above a salinity of 5 $^0/_{00}$ (Fig. 1).

The widespread appearance of MMA in Chesapeake Bay is unusual relative to previous studies in coastal and oceanic ecosystems and needs to be studied more closely. Similar concentrations were recently reported from a productive estuary in England.[48] Therefore, the production of MMA may be common to all highly productive estuaries. If so, the potential for impact from continuing pollutant loading is greater than currently thought. In addition, the controlling factor for MMA production needs to be identified.

The mechanism for the release of MMA, and its physiological significance relative to the previously

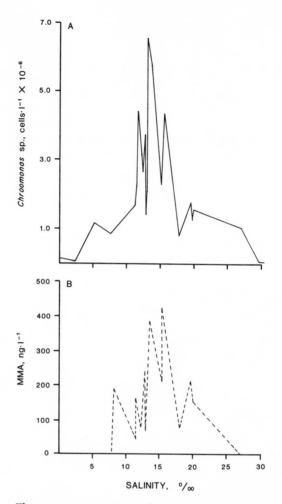

Fig. 1. The concentration of methylarsonate (MMA) and the density of <u>Chroomonas</u> sp. in samples taken along a Chesapeake Bay transect, July 1983. Taken from 19.

reported pathway of reduction to DMA is unknown.[46,49] The ability to methylate arsenate and release the products to the surrounding environment varies even within a particular species. Perhaps all algal species do not contain the necessary enzymes required for methylation of MMA. This mechanism remains to be determined.

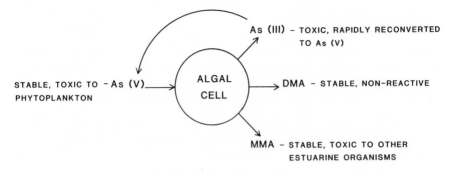

Fig. 2.   Summary of biologically-mediated arsenic transformations in productive aquatic systems.

Figure 2 summarizes the possible alterations of
arsenic speciation and toxicity mediated by phytoplankton.
Of particular importance is the production of methylated
compounds, with the production of DMA alleviating the
toxicity of arsenic to the ecosystem and the production
of MMA perhaps increasing arsenic toxicity to estuarine
fauna.

CHROMIUM BIOGEOCHEMISTRY AND TOXICITY

Chromium is an important industrial metal; it is used
in metallurgy, as an anticorrosive in cooling towers of
power plants, and as a tanning agent in leather manufac-
ture.[50]   In the hexavalent form it is mutagenic,[51]
carcinogenic,[52] and corrosive, as well as being potentially
toxic to the environment.[53]

Chromium exists in the environment in two valences,
+6 and +3. The hexavalent form is predicted to be the
stable form over most of the pH range and oxygen concentrations encountered in the environment; only at low pH or in
anoxic conditions is trivalent chromium predicted to be
the predominant form.[54] Analytical studies in oceanic and
estuarine environments show that while Cr(VI) is the
dominant form in oxygen-containing waters, much more
Cr(III) is often present than can be accounted for by
equilibrium with oxygen at the ambient pH.[55-57]

Hexavalent chromium is present in natural waters predominantly as the anion chromate (CrO_4^{-2}). This anion is similar to arsenate in that it is less subject to the usual mechanisms that control removal of trace metals from the water column (adsorption onto particles or phytoplankton uptake, followed by incorporation into larger particles and removal to the sediment).[10,58] However, Cr(III) is quite different from As(III) in that it is highly insoluble, exists in natural water largely as hydroxide complexes, and has a high affinity for organic and inorganic binding sites. In this respect, it more closely resembles iron.

The redox chemistry of chromium in natural waters is poorly understood at this time. Under laboratory conditions, both Cr(III) and Cr(VI) are kinetically inert. However, transformations in both directions have been observed in natural waters. Amdurer et al. have observed rapid ($t_{\frac{1}{2}} = 14$ d) oxidation of Cr(III) in Narragansett Bay water.[59] Conversely, north Pacific Ocean water containing low oxygen concentrations show much greater concentrations of Cr(III) than the overlying and underlying layers of higher oxygen content, indicating Cr(VI) reduction.[55] In seawater Cr(VI) can also be photoreduced by sunlight in the presence of natural dissolved organic compounds (Fig. 3).[60] The balance between Cr(III) and Cr(VI) in a given body of water is likely to be controlled by the input of the two forms, the effects of the largely unknown redox reactions, and the rates of loss of the two forms from the water rather than by simple chemical equilibrium with oxygen.

Hexavalent chromium is widely known to be the more toxic of the two valences of chromium to a wide variety of organisms, from phytoplankton to man.[53] In phytoplankton chromate acts as an analogue of the required sulfate ion; both toxicity and uptake of chromium are in proportion to the ratio of chromate to sulfate in the surrounding water.[10,61] Chromate enters the cell as an analogue of sulfate; the sulfate uptake system is apparently unable to exclude it due to its similar size, shape, and charge. Chromate appears to exert its toxic effect after entering the cell although the exact site of damage is unknown.

Because of the action of Cr(VI) toxicity through sulfate uptake, Cr(VI) toxicity to algae is highly

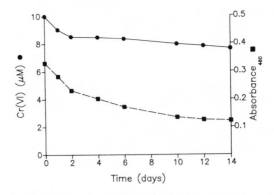

Fig. 3. Photoreduction of Cr(VI) in organic-rich seawater collected from Florida. Water shielded from light
or without dissolved organic compounds showed no loss of
Cr(VI).

dependent on the salinity of the water, the greatest
toxicity occurring at the lowest salinity.[61,62] In fresh
waters, where sulfate concentrations are highly variable
depending on the hydrological and geological conditions,
Cr(VI) toxicity is equally variable. In freshwaters of
below average sulfate concentration, chromate is toxic at
concentrations close to ambient water concentrations.

Although all algae surveyed to date show the effect
of sulfate on Cr(VI) toxicity, there are large differences
in the susceptibility of different species of algae to
Cr(VI), and there is evidence of a general taxonomic
grouping of resistance (Table 2). As with arsenic,
differential sensitivity to Cr(VI) could lead to alteration of phytoplankton species composition and species
succession in areas of chronic chromium pollution.

The extent to which phytoplankton alter the speciation of chromium after uptake is currently unknown.
Biochemical reactions alter chromium in different ways;
e.g., rat liver mitochondria readily take up Cr(VI) and
reduce it to Cr(III),[63] but barley seedlings take up
Cr(III) and oxidize it to Cr(VI).[64] Probably both
reactions can occur in different regions of any one cell,
mediated by different redox intermediates and enzymes.

Table 2. Toxicity of Cr(VI) to euryhaline phytoplankton, extrapolated to average freshwater and seawater.

| Taxonomic Class, Species | Clone | EC$_{50}$ (μM)[*] | |
		Freshwater	Seawater
Dinophyceae			
Prorocentrum mariae-lebouriae	PROML	0.11	25.0
Chrysophyceae			
Pseudoisochrysis sp.	TISO	0.11	26.5
Pavlova lutheri	MONO	0.13	30.0
Bacillariophyceae			
Thalassiosira weissflogii	ACTIN	0.14	32.4
Thalassiosira pseudonana	3H	0.19	44.2
Cyanophyceae			
Synechococcus sp.	SYN	0.67	152
Synechococcus bacillaris	UW405	1.26	287
Chlorophyceae			
Chlorella capsulata	∅.17	1.43	325
Chlorella sp.	FLAE	6.88	1560

[*]EC$_{50}$ is the concentration of Cr(VI) necessary to reduce algal growth rate by 50%.

Due to competition with sulfate, uptake of a large fraction of Cr(VI) from saline water is unlikely, and thus mass reduction of Cr(VI) by uptake and release as Cr(III) would be restricted to essentially freshwater environments. In addition, the high affinity between Cr(III) and proteins would suggest that it is unlikely that Cr(III) so produced would be released to the water. However, Cr(III) could be released from dead phytoplankton during decomposition and remineralization.

A more likely means by which phytoplankton could reduce Cr(VI) or Cr(III) in water would be by release of dissolved organic compounds (DOC). Natural DOC can rapidly reduce Cr(VI) in sunlight in the laboratory (Fig. 3); however, further studies are required to determine if this reaction is important in natural waters. It is also possible, but currently undemonstrated, that compounds released by phytoplankton may directly reduce Cr(VI) without requiring light.

The oxidation of Cr(III) to the more toxic Cr(VI) by phytoplankton is also possible, because this reaction has been observed in estuarine microcosms.[59] This oxidation would not be limited by salinity as would the reverse reaction because Cr(III) is not an analogue of a major seawater ion, as is Cr(VI). However, Cr(III) may be taken up by phytoplankton as an analogue of Fe(III), a necessary nutrient often in short supply in marine waters. Although the iron metabolism of phytoplankton has been studied in detail only recently, phytoplankton are able to accumulate iron even though ambient concentrations are extremely low.[65] Among the mechanisms employed by phytoplankton to capture iron is the excretion of siderophores, strong chelating agents for Fe(III).[66] The specificity of siderophores for Fe(III) over Cr(III) has not been investigated but experience with other similar analogue pairs suggests that such chelators might bind and transport significant amounts of the competing ion. If phytoplankton were to take up chromium in this way, some of it could be oxidized to Cr(VI) and released. Further research is needed to determine whether phytoplankton can oxidize Cr(III) as well as reduce Cr(VI).

Figure 4 summarizes chromium transformation in productive ecosystems, and the role of phytoplankton in these transformations. Unlike arsenic, the most important biologically-mediated process is the facilitation of redox shifts between Cr(III) and Cr(VI).

SUMMARY

Further study of the biological processes involved in controlling trace ion form, transport, and transformation in an estuarine water column will be necessary in future years. Although all inorganic elements will not react in

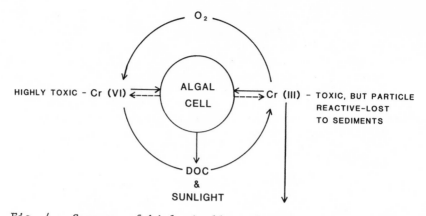

Fig. 4.   Summary of biologically-mediated chromium transformations in productive aquatic systems.

the same manner as the two discussed here, similar
geochemical and biological reactions will occur.

It is crucial that we understand how several
processes--physico-chemical controls of trace ion speciation, algal incorporation, transformation, and release,
and trace ion control of phytoplankton speciation--are
coupled in productive estuaries before we can predict
with confidence the impact of arsenic, chromium, or
similar toxic compounds within affected estuaries and
coastal oceans. In some instances the important effect
of phytoplankton uptake may be the production of toxic,
methylated compounds, because of their persistence,
stability, and toxicity to organisms within the ecosystem.
Arsenic, and perhaps selenium, tin, mercury, antimony,
and lead are elements of this type. In other cases
phytoplankton may reduce the toxicity of elements by
transforming them into less toxic forms. Reduction of
chromium and copper, either directly or by photoreduction,
or complexation (and detoxification) by DOC are examples
of this type of effect. In both instances, however, the
modifications mediated by phytoplankton are of significance
to the estuarine ecosystem. They must be considered when
impact assessments or predictions are made.

REFERENCES

1. WOOD, J.M. 1974. Biological cycles for toxic elements in the environment. Science (Wash., D.C.) 183: 1049-1052.
2. Environmental Protection Agency. 1983. Chesapeake Bay: A framework for action, 186 pp.
3. FÖRSTNER, U., G.T.W. WITTMANN. 1983. Metal Pollution in the Aquatic Environment. Springer-Verlag, Berlin, 486 pp.
4. LI, W.K.W. 1978. Kinetic analysis of interactive effects of cadmium and nitrate on growth of Thalassiosira fluviatilis (Bacillariophyceae). J. Phycol. 14: 454-460.
5. MOREL, N.M.L., J.G. RUETER, F.M.M. MOREL. 1978. Copper toxicity to Skeletonema costatum (Bacillariophyceae). J. Phycol. 14: 43-48.
6. RUETER, J.G. 1983. Effect of copper on growth, silicic acid uptake and soluble pools in the marine diatom Thalassiosira weissflogii. J. Phycol. 19: 101-104.
7. SUNDA, W.G., R.T. BARBER, S.A. HUNTSMAN. 1981. Phytoplankton growth in nutrient rich seawater: importance of copper-manganese cellular interactions. J. Mar. Res. 39: 567-586.
8. BLUM, J.J. 1966. Phosphate uptake by phosphate-starved Euglena. J. Gen. Physiol. 49: 1125-1136.
9. SANDERS, J.G., H.L. WINDOM. 1980. The uptake and reduction of arsenic species by marine algae. Estuarine Coastal Mar. Sci. 10: 555-567.
10. RIEDEL, G.F. 1985a. The relationship between Chromium(VI) uptake, sulfate uptake, and Chromium (VI) toxicity in the estuarine diatom Thalassiosira pseudonana. Aquat. Toxicol. 7: 191-204.
11. JUDSON, S. 1968. Erosion of the land. Am. Sci. 56: 356-374.
12. GORDON, G.E., W.H. ZOLLER. 1974. Study of the emissions from major air pollution sources and their atmospheric interactions. Annual report to NSF under Grant #GT-36338X.
13. MACKENZIE, F.T., R.J. LANTZY, V. PATERSON. 1979. Global trace metal cycles and predictions. Math. Geol. 11: 99-144.
14. CRECELIUS, E.A., M.H. BOTHNER, R. CARPENTER. 1975. Geochemistries of arsenic, antimony, mercury, and

related elements in sediments of Puget Sound.
Environ. Sci. Technol. 9: 325-333.

15. CARPENTER, R., M.L. PETERSON, R.A. JAHNKE. 1978.
Sources, sinks, and cycling of arsenic in the
Puget Sound region. In Estuarine Interactions.
(M.L. Wiley, ed.), Academic Press, New York, pp.
459-480.

16. ANDREAE, M.O., J.T. BYRD, P.N. FROELICH. 1983.
Arsenic, antimony, germanium, and tin in the Tejo
estuary, Portugal: modeling a polluted estuary.
Environ. Sci. Technol. 17: 731-737.

17. LANGSTON, W.J. 1980. Arsenic in U.K. estuarine
sediments and its availability to benthic organisms.
J. Mar. Biol. Assoc. U.K. 60: 869-881.

18. KNOX, S., W.J. LANGSTON, M. WHITFIELD, D.R. TURNER,
M.I. LIDDICOAT. 1984. Statistical analysis of
estuarine profiles: II Application to arsenic in
the Tamar estuary (S.W. England). Estuarine
Coastal Shelf Sci. 18: 623-638.

19. SANDERS, J.G. 1985. Arsenic geochemistry in
Chesapeake Bay: dependence upon anthropogenic
inputs and phytoplankton species composition.
Mar. Chem. 17: 329-340.

20. ANDREAE, M.O. 1978. Distribution and speciation of
arsenic in natural waters and some marine algae.
Deep-Sea Res. 25: 391-402.

21. WASLENCHUK, D.G. 1978. The budget and geochemistry
of arsenic in a continental shelf environment.
Mar. Chem. 7: 39-52.

22. SANDERS, J.G. 1980. Arsenic cycling in marine
systems. Mar. Environ. Res. 3: 257-266.

23. FERGUSON, J.F., J. GAVIS. 1972. A review of the
arsenic cycle in natural waters. Water Res. 6:
1259-1274.

24. LOWENTHAL, D.H., M.E.Q. PILSON, R.H. BYRNE. 1977.
The determination of the apparent dissociation
constants of arsenic acid in seawater. J. Mar.
Res. 35: 653-669.

25. BRAMAN, R.S. 1975. Arsenic in the environment.
In Arsenical Pesticides. (E.A. Woolson, ed.),
American Chemical Society, Washington, D.C.,
pp. 108-123.

26. SANDERS, J.G. 1983. Role of marine phytoplankton
in determining the chemical speciation and biogeo-
chemical cycling of arsenic. Can. J. Fish. Aquat.
Sci. 40 (Suppl. 2): 192-196.

27. SANDERS, J.G. 1986. Alteration of arsenic transport and reactivity in coastal marine systems due to biological transformation. Rapp. P.-v. Reun. Cons. int. Explor. Mer. 186: 185-192.

28. GUILLARD, R.R.L., P. KILHAM. 1977. The ecology of marine planktonic diatoms. In The Biology of Diatoms. (D. Werner, ed.), Blackwell Sci. Publ., Oxford, pp. 372-469.

29. KILHAM, P., S.S. KILHAM. 1980. The evolutionary ecology of phytoplankton. In The Physiological Ecology of Phytoplankton. (I. Morris, ed.), Univ. of California Press, Berkeley, pp. 571-597.

30. MORRIS, R.J., M.J. McCARTNEY, A.G. HOWARD, M.H. ARBAB-ZAVAR, J.S. DAVIS. 1984. The ability of a field population of diatoms to discriminate between phosphate and arsenate. Mar. Chem. 14: 259-265.

31. BUDD, K., S.R. CRAIG. 1981. Resistance to arsenate toxicity in the blue-green alga Synechococcus leopoliensis. Can. J. Bot. 59: 1518-1521.

32. PEOPLES, S.A. 1975. Review of arsenical pesticides. In E.A. Woolson, ed., op. cit. Reference 25, pp. 1-12.

33. KNOWLES, F.C. 1982. The enzyme inhibitory form of inorganic arsenic. Biochem. International 4: 647-653.

34. KNOWLES, F.C., A.A. BENSON. 1983. The biochemistry of arsenic. Trends Biochem. Sci. 8: 178-180.

35. JOHNSON, D.L., M.E.Q. PILSON. 1975. The oxidation of arsenite in seawater. Environ. Lett. 8: 157-171.

36. SANDERS, J.G. 1978. Interaction of Arsenic Species and Marine Algae. Ph.D. Thesis, University of North Carolina, Chapel Hill.

37. SCUDLARK, J.R., D.L. JOHNSON. 1982. Biological oxidation of arsenite in seawater. Estuarine Coastal Shelf Sci. 14: 693-706.

38. LUNDE, G. 1977. Occurrence and transformation of arsenic in the marine environment. Environ. Health Perspect. 19: 47-52.

39. ANDREAE, M.O. 1979. Arsenic speciation in seawater and interstitial waters: The influence of biological-chemical interactions on the chemistry of a trace element. Limnol. Oceanogr. 24: 440-452.

40. PLANAS, D., F.P. HEALEY. 1978. Effects of arsenate on growth and phosphorus metabolism of phytoplankton. J. Phycol. 14: 337-341.

41. SANDERS, J.G. 1979. Effects of arsenic speciation and phosphate concentration on arsenic inhibition of Skeletonema costatum (Bacillariophyceae). J. Phycol. 15: 424–428.

42. SANDERS, J.G., P.S. VERMERSCH. 1982. Response of marine phytoplankton to low levels of arsenate. J. Plankton Res. 4: 881–893.

43. SANDERS, J.G. 1982. Adaptive behavior of euryhaline phytoplankton to stress: response to chronic, low-level additions of trace metals. Semiannual progress summary to NOAA under Grant #NOAA-NA81RAD00032.

44. SANDERS, J.G., S.J. CIBIK. 1985. Adaptive behavior of euryhaline phytoplankton communities to arsenic stress. Mar. Ecol. Prog. Ser. 22: 199–205.

45. ANDREAE, M.O., D.K. KLUMPP. 1979. Biosynthesis and release of organo-arsenic compounds by marine algae. Environ. Sci. Technol. 13: 738–741.

46. NISSEN, P., A.A. BENSON. 1982. Arsenic metabolism in freshwater and terrestrial plants. Physiol. Plant, 54: 446–450.

47. JOHNSON, D.L., R.M. BURKE. 1978. Biological mediation of chemical speciation II. Arsenate reduction during marine phytoplankton blooms. Chemosphere 8: 645–648.

48. HOWARD, A.G., M.H. ARBAB-ZAVAR, S. APTE. 1984. The behavior of dissolved arsenic in the estuary of the river Beaulieu. Estuarine Coastal Shelf Sci. 19: 493–504.

49. CHALLENGER, F. 1945. Biological methylation. Chem. Rev. 36: 315–361.

50. MORNING, J.L. 1978. Chromium. In Minerals Handbook 1976, Vol. I, Bureau of Mines, U.S. Government Printing Office, Washington, D.C., pp. 297–308.

51. ARLAUSKAS, A., R.S.U. BAKER, A.M. BONIN, R.K. TANDON, P.T. CRISP, J. ELLIS. 1985. Mutagenicity of metal ions in bacteria. Environ. Res. 36: 379–388.

52. ENTERLINE, P.E. 1974. Respiratory cancer among chromate workers. J. Occup. Med. 16: 523–526.

53. TOWILL, L.E., C.R. SHRINER, J.S. DRURY, A.S. HAMMONS, J.W. HOLLEMAN. 1978. Reviews of environmental effects of pollutants. III. Chromium. EPA – 600/1/78/023 U.S. E.P.A. Health Effects Research Laboratory, Cincinnati, Ohio.

54. ELDERFIELD, H. 1970. Chromium speciation in seawater. Earth Planet. Sci. Lett. 9: 10–16.

55. CRANSTON, R.E., J.W. MURRAY. 1978. The determination
    of chromium species in natural waters. Anal. Chim.
    Acta 99: 275-282.
56. CRANSTON, R.E. 1983. Chromium in Cascadia Basin,
    Northeast Pacific Ocean. Mar. Chem. 13: 109-125.
57. JEANDEL, C., J.-F. MINSTER. 1984. Isotope dilution
    measurement of inorganic chromium(III) and total
    chromium in seawater. Mar. Chem. 14: 347-364.
58. MURRAY, J.W., B. SPELL, B. PAUL. 1983. The contrasting
    geochemistry of manganese and chromium in the
    eastern tropical Pacific Ocean. In Trace Metals
    in Seawater. (C.S. Wong, E. Boyle, K.W. Bruland,
    J.D. Burton, E.D. Goldberg, eds.), Plenum Press,
    New York, pp. 643-670.
59. AMDURER, M., D. ADLER, P.H. SANTSCHI. 1983. Studies
    of the chemical forms of trace elements in
    seawater using radiotracers, ibid., pp. 537-562.
60. RIEDEL, G.F. 1985b. Photoreduction of Cr(VI) in
    natural waters. Trans. Am. Geophys. Union 66:
    1257.
61. RIEDEL, G.F. 1984. Influence of salinity and sulfate
    on the toxicity of chromium (VI) to the estuarine
    diatom Thalassiosira pseudonana. J. Phycol. 20:
    496-500.
62. FREY, B.F., G.F. RIEDEL, A.E. BASS, L.F. SMALL. 1983.
    Sensitivity of estuarine phytoplankton to hexa-
    valent chromium. Estuarine Coastal Shelf Sci. 17:
    181-187.
63. ALEXANDER, J., J. AASETH, T. NORSETH. 1982. Uptake
    of chromium by rat liver mitochondria. Toxicology
    24: 115-122.
64. SKEFFINGTON, R.A., P.R. SHREWRY, P.J. PETERSON. 1976.
    Chromium uptake and transport in barley seedlings
    (Hordeum vulgare L.). Planta 132: 209-214.
65. ANDERSON, M.A., F.M.M. MOREL. 1982. The influence
    of aqueous iron chemistry on the uptake of iron
    by the coastal diatom Thalassiosira pseudonana.
    Limnol. Oceanogr. 27: 798-813.
66. TRICK, C.G., R.J. ANDERSON, A. GILLANI, P.J. HARRISON.
    1983. Prorocentrin: An extracellular siderophore
    produced by the marine dinoflagellate Prorocentrum
    minimum. Science (Wash., D.C.) 219: 306-309.

Chapter Six

PLANT AND BACTERIAL CYTOCHROMES P-450:   INVOLVEMENT IN
HERBICIDE METABOLISM

DANIEL P. O'KEEFE, JAMES A. ROMESSER
AND KENNETH J. LETO

Central Research and Development Department
E.I. Du Pont De Nemours and Company Inc
Experimental Station
Wilmington, Delaware 19898

INTRODUCTION

Although herbicides have been used in weed control
for many years, our understanding of the biological
chemistry involved in their metabolism and degradation is
still in its infancy. These reactions take place both in
plants which have been treated with herbicides, and in
soil microorganisms.[1-3] Metabolism in a soil microorganism
can result in increased water solubility of the herbicide,
with corresponding increased bioavailability and
biodegradability ultimately leading to the complete
degradation of the herbicide. Herbicide metabolism in
plants can result in altered phytotoxicity providing a
mechanism of crop selectivity based on the differential
metabolism of a herbicide in crop plants as compared with
weeds. This very common mechanism of selectivity, which
relies largely on uncharacterized enzymatic pathways, is

151

in stark contrast to the alternative and less common mechanism of crop selectivity in which the characteristics of
the target site enzyme for the herbicide[4-7] are modified
to a herbicide resistant form.[8-12]

While it is difficult to categorize completely the
types of reactions involved in the degradation of herbicides
by plants and microorganisms, some generalizations can be
made. Microorganisms rely heavily on hydrolytic enzymes
(e.g., esterases, amidases),[2] while plants are more prone
to utilize hydroxylation and subsequent glycosylation or
conjugation with a glutathione-S-transferase type
system.[1,13-17] The aryl and alkyl hydroxylations (and
their mechanistically similar counterparts, O-dealkylation,
and N-dealkylation) are carried out by cytochrome P-450
dependent monooxygenases and are well characterized in a
number of animals and some bacteria. The study of plant
P-450 dependent monooxygenases has been much more
restricted to limited reports of only partially characterized cytochromes P-450 capable of metabolizing
herbicides.[1,14-16,18,19] In this chapter, we will confine
our discussion to the metabolism of herbicides by cytochrome
P-450 dependent monooxygenases, and to the characterization
of plant cytochromes P-450 in general. We will describe a
number of approaches useful in characterizing plant
cytochromes P-450 and will define some of the fundamental
problems in the study of these plant proteins. Finally,
we will describe the cytochrome P-450 dependent sulfonylurea
monooxygenase from Streptomyces griseolus to illustrate the
essential features of an efficient and biochemically well
characterized system that metabolizes a herbicide.

CHARACTERISTICS OF CYTOCHROME P-450 MONOOXYGENASES

It is helpful to outline the characteristics of
cytochrome P-450 dependent monooxygenases. Cytochromes
P-450 are heme proteins in which one of the axial
heme ligands is a cysteine sulfhydryl provided by
the protein. It is this sulfhydryl ligand which contributes
most strongly to the unique visible spectroscopic properties
of the ferrocytochrome:CO complex which absorbs at about
450 nm[20] (see also Fig. 4). This absorption is sufficiently

unique to distinguish cytochrome P-450 from most other cytochromes, although some interferences are possible.[21]

The P-450 dependent monooxygenase system requires O_2 and reduced pyridine nucleotide and is inhibited by carbon monoxide. The xenobiotic metabolizing systems of mammalian liver utilize a flavoprotein NADPH:cytochrome P-450 reductase. Another b-cytochrome, cytochrome b_5, is usually found in association with the cytochrome P-450 but is not always directly involved in the transfer of reducing equivalents.[22] Cytochromes P-450 from adrenal gland mitochondria utilize the flavoprotein adrenodoxin reductase and the iron-sulfur protein (2Fe-2S) adrenodoxin in a sequential transfer of reducing equivalents to cytochrome(s) P-450.[22] All of these mammalian cytochromes P-450, associated reductases, and cytochrome b are intrinsic membrane proteins. In contrast, the best characterized bacterial cytochrome P-450 monooxygenase system is the camphor-5-exo-hydroxylase from Pseudomonas putida. This is a soluble protein system composed of a flavoprotein reductase (putidaredoxin reductase), an iron-sulfur protein (putidaredoxin, 2Fe-2S), and a cytochrome P-450 designated as cytochrome P-450$_{CAM}$.[23,24]

The reaction mechanism of cytochrome P-450, a monooxygenase, involves the reduction of molecular oxygen to one molecule of water and one six electron oxygen atom ("oxene") bound to the heme iron.[25] Subsequent reactions of this oxene involve a variety of possible mechanisms,[24,26,27] the result being a hydroxylated substrate (or O-, or N-dealkylated substrate by subsequent nonenzymatic rearrangement).

These characteristics of cytochrome P-450 dependent monooxygenases dictate the following criteria for determining the involvement of a cytochrome P-450 in an enzymatic reaction: a) type of reaction products; b) reaction requirements for reducing equivalents, O_2, and inhibition by CO; c) spectral evidence for both the existence of, and substrate binding to,[28] cytochrome P-450; and d) purification of catalytically active cytochrome P-450. We will apply these criteria to the available information on the metabolism of herbicides by plants in order to evaluate more critically the role of cytochromes P-450 in this area.

PLANT XENOBIOTIC METABOLISM AND CYTOCHROMES P-450

There are an increasing number of enzymatic reactions
in plants which have been shown to be catalyzed by microsomal cytochrome P-450 dependent enzymes.[18,19,29-39]
These enzymes are often involved in the biosynthesis of
plant secondary metabolites, and the best characterized
of the plant P-450's is the trans-cinnamic acid
4-hydroxylase which is found in a wide variety of plant
tissues.[29,34,37-39] The references cited also describe
cytochromes P-450 which are apparently involved in the
metabolism of xenobiotics; these include the monuron
N-demethylase,[18,19] p-chloro-N-methylaniline
demethylase,[31,32] and aldrin epoxidase.[33] Frear and
co-workers in 1969 reported the N-demethylation of the
herbicide monuron in a microsomal fraction from etiolated
cotton hypocotyls. This activity required reduced
pyridine nucleotide, molecular oxygen, and was inhibited
by carbon monoxide.[18] Cytochrome P-450 was not initially
detected, although in a later publication a demonstrable
amount of cytochrome P-450 was observed.[19] More recently
Sweetser,[14] and Erbes[15] have speculated on the existence
of an inducible cytochrome P-450 dependent monooxygenase
involved in the hydroxylation of chlorsulfuron by wheat.
This was based on the incorporation of one atom of ^{18}O
from $^{18}O_2$ into a hydroxylated chlorsulfuron metabolite
and the sensitivity of chlorsulfuron metabolism to
inhibitors of protein biosynthesis.

PROPERTIES OF OTHER KNOWN PLANT CYTOCHROMES P-450

While the mammalian literature contains a wealth of
information concerning the purification and characterization
of a large number of hepatic cytochromes P-450,[40,41,47-49]
almost no plant P-450s have been studied in similar detail.
We outline here the information available on some of the
best characterized plant cytochromes P-450, and the
techniques commonly employed to study these cytochromes.

At the molecular level perhaps the best studied plant
cytochromes P-450 are those found in microsomes obtained
from tulip bulbs and water aged tubers of Jerusalem artichoke. A cytochrome P-450 from tulip bulb microsomes has
been isolated by solubilization in the nonionic detergent
Emulgen 911, followed by ion exchange chromatography.[42,43]

Two additional (minor) P-450 forms were detected, and were
claimed to be immunochemically distinct.[42] Although
tulip bulbs contain large amounts of cytochrome P-450,
its physiological role is unknown. The cinnamic acid
hydroxylase of Jerusalem artichoke depends on a cytochrome
P-450 which has also been recently purified in an enzymatically active form.[44] This purification depended on
solubilization in the zwitterionic detergent CHAPS, and
subsequent ion exchange chromatography. Minor fractions
containing cytochrome P-450 were detected during this
purification procedure; however it is not clear if these
represent isozymes distinct from the major cytochrome
P-450. Jerusalem artichoke microsomes were previously
reported to contain a second cytochrome P-450 associated
with lauric acid hydroxylase activity.[30] Notably, the
cinnamic acid hydroxylase of Jerusalem artichoke is
inducible. Slicing the tissue and letting it sit for
several hours results in the appearance of enzymatic
activity and increased cytochrome P-450 levels. The
presence of additional compounds during the water aging
process (e.g., $MnCl_2$ or phenobarbital, ethanol, and some
herbicides) can result in even greater levels of enzyme.[45-46]

Solubilization with detergents is one of the first
steps in the purification of any intrinsic membrane
protein. The two successful purifications listed above
used either Emulgen 911 in combination with sodium cholate,
or CHAPS in combination with Emulgen 911. We have found
that Triton X-100 substitutes quite well for Emulgen 911,
although CHAPS may be more effective because of its much
lower critical micelle concentration. With both, we have
obtained >90% solubilization of tulip bulb or Jerusalem
artichoke cytochrome P-450, and with CHAPS >85% solubilization of wheat cytochrome P-450.

Anion exchange HPLC has proven to be an extremely
powerful tool in the analysis of detergent solubilized
cytochromes P-450 from mammalian liver. Using this
technique it is possible to separate chromatographically
most of the detergent-solubilized hepatic cytochromes
P-450.[47-49] In Figure 1 we have analyzed detergent-
solubilized extracts from a variety of plant tissue
sources. Although the chromatograms represent detection
at a single wavelength, spectral analysis of the chromatogram for the characteristic Soret absorption is possible
using a diode array detector and enables a tentative

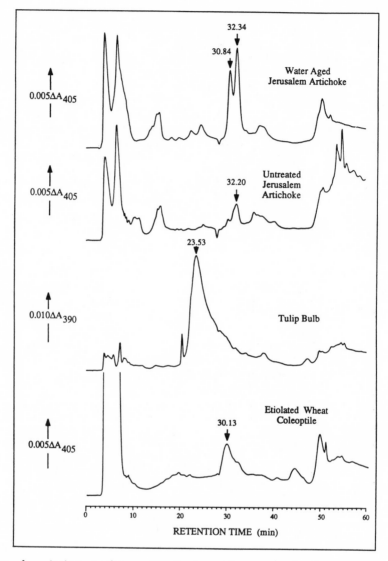

Fig. 1. Anion exchange HPLC chromatograms of plant microsomal proteins solubilized by Triton X-100. Chromatography was performed on a TSK-DEAE-3SW column, at a flow rate of 0.75 ml/min, with a linear gradient of 0 to 30% solvent B from 5 to 45 min, and 30% to 100% B from 45 to 50min. Solvent A: 20 mM Tris-Acetate, pH 7.0, 0.2% Triton X-100; Solvent B: Solvent A + 0.8 M Sodium Acetate. Approximately 1 mg of solubilized protein was injected.

identification of a peak as a cytochrome. If sufficient
sample is available, collection of a fraction and subsequent
analysis for cytochrome P-450 content can confirm the
identity of a suspected peak. For example, the 23.53
minute peak in the tulip bulb preparation has been
identified as a cytochrome P-450. Figure 1 compares the
chromatograms from four plant microsomal extracts
solubilized with Triton X-100. In general those peaks
in the 3-8 min region and 48-60 min are due to chromophores
other than cytochrome. The induction of two heme-containing
proteins in the water aged Jerusalem artichoke preparations
is clearly seen when compared to control tissue. The
presence of a single major cytochrome P-450 peak (albeit
broad) in tulip bulbs, and the presence of a single heme-
containing peak in the etiolated wheat coleoptiles is also
observed. The "cytochrome b_5" of plant microsomes is not
effectively solubilized by Triton X-100 under the conditions
used here, and in the Jerusalem artichoke preparation used
in this experiment, aging of slices in water did not result
in more than 2-fold increase in the content of this
cytochrome.

Another approach useful in the analysis of plant
cytochromes in general is the use of techniques to
visualize heme-containing polypeptide bands on electrophoresis gels. This technique relies upon the heme-catalyzed
peroxidase activity visualized with tetramethylbenzidine
and was originally developed for the detection of hepatic
cytochromes P-450.[50] Either native or denaturing conditions
may be used; however the analysis of microsomal fractions
requires the use of at least mild denaturing conditions to
insure adequate electrophoretic resolution. The potential
for dissociation of the heme from the apoprotein, especially
with b-type cytochromes (such as P-450's), is the major
complication encountered with this technique. Figure 2
compares microsomal proteins from water-aged tissue of
Jerusalem artichoke (lanes labeled 1) and control tissue
(lanes labeled 2) visualized with both a heme stain and a
general protein stain (Coomassie brilliant blue). Although
there are a number of changes in the overall protein profile,
particularly in the 50-100 kD region, there are only 3
noticeable increases of heme-associated bands which appear
in the water aged tissue. The molecular weight of the
purified cinnamic acid hydroxylase cytochrome P-450 of
Jerusalem artichoke is 56 kD, and so it seems likely that
the band at 63 kD is this cytochrome. The bands at 33 and

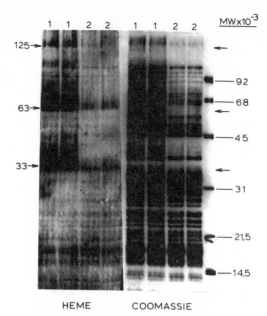

HEME   COOMASSIE

Fig. 2. LDS Gel electrophoresis of microsomal proteins
from Jerusalem artichoke. Lane 1: water aged Jerusalem
artichoke tubers. Lane 2: untreated Jerusalem artichoke
tubers.

125 kD are more difficult to interpret and at this time
their identity is unclear.

A third approach of potential utility in the detection
and analysis of plant cytochromes P-450 is to use immuno-
logical techniques to search microsomal extracts for
proteins antigenically related to known P-450s. An obvious
requirement here is that one have a purified protein of
interest (e.g., a cytochrome P-450 from a related tissue
source, or one catalyzing a relevant activity) against
which antibodies can be raised. So far, this sort of
"heterologous probing" has not been successful in identi-
fying plant cytochromes P-450. In one case, antibodies
raised against tulip bulb cytochrome P-450 failed to
cross-react with P-450 from Jerusalem artichoke, cauliflower
(buds or leaves), avocado mesocarp, potato tubers, or even
with P-450 from similar plant tissue: bulbs from lilies,

allium, narcissus or gladiolus.[42] As shown in Table 1, an antibody raised against a bacterial cytochrome P-450 involved in the metabolism of sulfonylurea herbicides fails to identify any cross-reactive proteins in a wide variety of P-450 containing plant tissues, including wheat and corn, where sulfonylurea metabolism is thought to be mediated by cytochrome P-450.[14,15] The antibodies in this limited sampling are excellent probes for studying cytochromes P-450 in their respective systems, but have not lead to the identification of novel P-450s in plants. One way around the problem of narrow antibody specificity may be to amplify the number of antigenic determinants discreetly sampled by generating monoclonal antibodies against P-450s of interest and testing a battery of them for cross-reactivity to plant microsomal extracts.[51]

We have shown in this, and in the previous section, that there is a wide variety of plant cytochrome P-450 dependent monooxygenases functional both for xenobiotic metabolism and plant secondary metabolism. In order to observe this spectrum of activities it is necessary to look at a wide variety of plant species and a number of plant tissue types.[1,14,15,18,19,29-39] In some cases the substrate specificity is high,[39] whereas in other cases the specificity is broader.[31,32] It is also evident that cytochromes P-450 from the same tissue of related plant species are immunologically distinct,[42] and the anion exchange properties of cytochromes P-450 from a small sampling of plant tissues suggests that each one contains no more than two predominant cytochromes P-450. This is in contrast to the mammalian liver microsomal system which is composed of at least 8 well characterized major forms of cytochrome P-450 which vary in substrate specificity, and in the spectra of inducibility.[40,41] One might argue that the autotrophic plants simply have less need for such a broad spectrum of proteins specialized in the detoxification of xenobiotics. Alternatively, Hendry has suggested that the detoxifying cytochromes P-450 are lost upon the vacuolation of maturing plant tissues at the same time that the ability to accumulate potential toxins in vacuoles is acquired.[52] Nonetheless, considering the plant kingdom as a whole, the detoxifying ability seems quite high,[1] and so the diversity of xenobiotic metabolizing capability by plant cytochrome P-450 may reside more in the variety of cytochromes P-450 found in different plant species and

Table 1. Detection of proteins which cross-react with
cytochrome P-450 SU1 from Streptomyces griseolus

Sample	Cross Reactivity (Immunoblot)
Purified Cytochrome P-450, Procaryotic	
Cytochrome P-450$_{SU1}$ (S. griseolus)	+++
Cytochrome P-450$_{SU2}$ (S. griseolus)	weak
Cytochrome P-450$_{CON}$ (S. griseolus)	very weak
Cytochrome P-450$_{CAM}$ (Ps. putida)	none
Chloroplasts	
Cytochromes b$_6$, b$_{559}$, f (maize, pea thylakoids)	none
Stroma (wheat)	none
Other Heme Proteins	
Catalase (bovine liver)	none
Horseradish peroxidase	none
Cytochrome c (horse heart)	none
Plant Microsomes Containing Cytochrome P-450	
Jerusalem artichoke (cinnamic acid hydroxylase)	none
Tulip bulb	none
Avocado (p-chloro-N-methylaniline demethylase)	none
Etiolated wheat coleoptiles	none

Crossreactivities were determined by immunodecoration of
nitrocellulose (Western blots) using rabbit antiserum cross-
reactive to authentic cytochrome P-450 SU1 followed by goat
anti-rabbit horseradish peroxidase conjugated second
antibody (IgG).

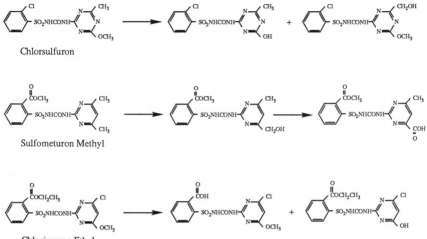

Chlorsulfuron

Sulfometuron Methyl

Chlorimuron Ethyl

Fig. 3. Metabolites of sulfonylurea herbicides in Streptomyces griseolus

different "immature" tissues than in the ability of any one plant tissue to metabolize a variety of xenobiotics.

THE SULFONYLUREA MONOOXYGENASE OF STREPTOMYCES GRISEOLUS

One experimental approach which effectively bypasses the problems associated with the study of cytochromes P-450 in plants is to use bacteria which perform the same, or analogous, metabolic transformations of the herbicide. The soil actinomycete, Streptomyces griseolus, perform a variety of co-metabolic transformations on several sulfonylurea herbicides (see Fig. 3).[3,16] In the case of chlorsulfuron, the hydroxymethyl metabolite formed is identical to one of the metabolites of chlorsulfuron in corn.[14] Figure 3 also illustrates the co-metabolism of two other sulfonylurea herbicides, sulfometuron methyl, and chlorimuron ethyl by S. griseolus. Except for the transient accumulation of an intermediate in the metabolism of sulfometuron methyl, the metabolites are all end products which accumulate in the culture medium. One can infer from the types of reactions carried out (i.e., the hydroxylation of all three compounds

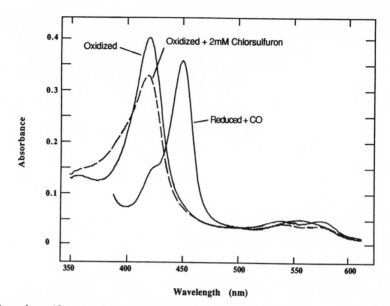

Fig. 4. Absorption spectra of purified cytochrome P-450$_{SU1}$ from <u>Streptomyces</u> <u>griseolus</u>

and the O-demethylation of chlorsulfuron and chlorimuron ethyl) that these reactions are carried out by a monooxygenase. Indeed the requirements for cell-free activity (NADPH or NADH, O_2, inhibition by CO) are consistent with the presence of one or more cytochrome P-450 dependent monooxygenases in <u>S. griseolus</u>. As described in a previous section, the involvement of a cytochrome P-450 is further indicated by the observation of a cytochrome P-450 absorption spectrum, which undergoes substrate dependent changes.[16] Figure 4 shows the absorption spectra of various states of purified cytochrome P-450$_{SU1}$ from <u>S. griseolus</u>. The absorption of the dithionite reduced cytochrome in the presence of carbon monoxide is centered at 449 nm. The absorption maximum for the oxidized cytochrome at 418 nm is typical for a cytochrome P-450 that is almost entirely low spin.[28] The change in the spectrum of the oxidized cytochrome which occurs after addition of the substrate, chlorsulfuron, provides evidence that the cytochrome actually binds the sulfonylurea. This absorption change (decrease at 418 nm, increase at 390 nm) is characteristic of a shift in the spin

Table 2. Sulfometuron methyl hydroxylase activity and cytochrome P-450 concentration in cell-free extracts of Streptomyces griseolus induced with sulfonylureas.

Inducer	Hydroxylase Activity (nmol/min/mg protein)	Cytochrome P-450 (pmol/mg protein)
None added	0	57
Chlorsulfuron	4.1	206
Sulfometuron methyl	4.2	328
Chlorimuron ethyl	7.8	431

Sulfonylurea inducer was added to growing cells 16 h prior to harvest. Sulfometuron methyl hydroxylase activity was measured in 35,000 x g cell-free extracts containing 0.05 M MOPS buffer, pH 7.2, 400 µM NADH, and 300 µM sulfometuron methyl. Cytochrome P-450 was measured in 100,000 x g cell-free extracts.

equilibrium of the cytochrome from predominantly low spin to partially high spin. Since the low spin form (S=1/2) absorbs maximally at about 418 nm, and the high spin form (S=5/2) absorbs maximally at about 390 nm, any change in the equilibrium levels of low or high spin (often the result of substrate binding) results in a corresponding shift in the absorption spectrum.

Cytochromes P-450 in animal systems are characteristically induced by their substrates,[40] aiding in the identification of proteins likely to be the cytochrome P-450 (see also the example of Jerusalem artichoke). Consistent with this, the data in Table 2 shows that cell-free extracts of S. griseolus, grown in the presence of a variety of sulfonylurea herbicides, exhibit an increase in the level of sulfometuron methyl hydroxylase activity, and a corresponding increase in the level of total cytochrome P-450. Both the increased rates of metabolism, and the increased levels of cytochrome P-450 are inhibited by the protein biosynthesis inhibitor, chloramphenicol. These results suggest that there is de novo synthesis of cytochrome(s) P-450 in response to

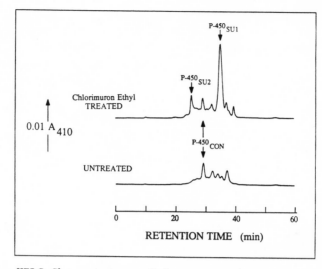

Fig. 5. HPLC Chromatogram of Streptomyces griseolus cell extracts. Chromatography was performed on a TSK DEAE-3SW column at a flow rate of 0.75 ml/min using a linear gradient of 0 to 100% solvent B from 1 to 40 minutes. Solvent A: 20 mM Tris-acetate, pH 7.0; Solvent B Solvent A + 0.8 M sodium acetate.

the sulfonylurea, and that these new cytochromes are responsible for the observed increase in enzymatic activity.[16] The anion exchange HPLC chromatogram in Figure 5 supports this conclusion. Here it is evident that there is one form of cytochrome P-450 which is present in extracts of cells grown in the presence or in the absence of sulfonylurea. This appears to be a constitutive protein and is designated cytochrome P-450$_{CON}$. Additionally, there are two cytochromes P-450 which are formed in response to the presence of the sulfonylurea inducer, chlorimuron ethyl: a major form, designated cytochrome P-450$_{SU1}$, and a minor form, designated cytochrome P-450$_{SU2}$.

We have taken advantage of the inducibility of cytochrome P-450$_{SU1}$ in S. griseolus to identify and isolate this protein from polyacrylamide gels for use as an immunogen to raise polyclonal antibodies. As shown in Table 1 we have obtained an antibody that is highly

selective for the SU1 isozyme. Immunoblotting experiments
indicate that cytochrome P-450$_{SU1}$ is immunologically
distinct from the SU2 and CON isozymes in S. griseolus,
from the "procaryotic" b-type cytochromes found in higher
plant chloroplasts, and, significantly, from the purified
camphor hydroxylase obtained from Ps. putida. This antibody also fails to cross-react with any microsomal
cytochromes P-450 from wheat, a plant which metabolizes
chlorsulfuron through a pathway involving cytochrome
P-450,[14,15] nor from the well studied cytochrome P-450
systems in Jerusalem artichoke, tulip bulb, or avocado
plant. We have exploited the specificity and high degree
of sensitivity of this antibody to study the induction of
cytochrome P-450$_{SU1}$ in S. griseolus under a variety of
conditions.

We have also purified native, catalytically active
cytochromes P-450$_{SU1}$ and P-450$_{SU2}$ to homogeneity, and
have obtained a partially purified preparation of cytochrome P-450$_{CON}$. With these preparations we have been
able to reconstitute an enzymatically active system. The
results of these experiments indicate that the complete
sulfonylurea monooxygenase system of S. griseolus consists
of either cytochrome P-450$_{SU1}$ or P-450$_{SU2}$, together with
an iron-sulfur protein and an NAD(P)H:iron-sulfur protein
reductase. The iron-sulfur protein is unique from other
iron sulfur clusters involved in cytochrome P-450 systems
in that it contains approximately 4 irons and 4 sulfurs,
although it is not clear if this protein contains a single
4Fe-4S cluster, or some combination of 1Fe, 2Fe-2S, or
3Fe-3S clusters. The reductase is still poorly characterized, but has about equal affinity for NADPH or NADH.
The two flavoprotein reductases, putidaredoxin reductase
from Ps. putida, or ferredoxin:NADP reductase from spinach
chloroplasts, substitute equally well in the system.

The overall organization of the three cytochromes
P-450 and the reductase system of S. griseolus is shown
in Figure 6. Cytochrome P-450$_{SU1}$ is capable of carrying out
two reactions on the sulfonylurea chlorimuron ethyl: an
O-demethylation, and a de-esterification. The two reactions are mutually exclusive, and the products are formed
in the ratio of 1:1 approximately. The mechanism of
de-esterification by cytochrome P-450$_{SU1}$ is not known,
although the reaction requires reducing equivalents and

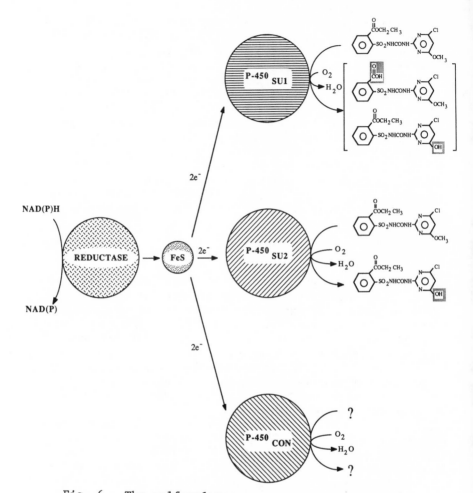

Fig. 6. The sulfonylurea monooxygenase system of
Streptomyces griseolus

takes place in preparations containing only purified
cytochrome P-450$_{SU1}$, and reductase components. We have
not analyzed for the other product(s) of this reaction;
however, it seems probable that this reaction is an
oxidative de-esterification mediated by cytochrome P-450.
This point is worth emphasizing since de-esterification
is such a common mechanism for degradation of xenobiotics
and it is usually assumed that a hydrolytic enzyme is

involved. Cytochrome P-450$_{SU2}$ carried out only the
O-demethylation of chlorimuron ethyl, and thus represents a
more specific isozyme (in terms of the products formed) of
the sulfonylurea monooxygenase. Cytochrome P-450$_{CON}$ has
an unknown function in the cell but because it is always
present (under the growth conditions used here) it would
appear that its function is essential. Furthermore, the
ability of purified cytochrome P-450$_{SU1}$ to confer the
capability of hydroxylating sulfonylurea onto untreated
cell-free extracts suggests the levels of the reductase
and iron sulfur protein are unaffected by the sulfonylurea
inducers. So it is important to note that cells of $\underline{S}$.
griseolus need to make only one protein to render them
capable of sulfonylurea monooxygenase reactions, since
the rest of the reductase system is already present.

SUMMARY

We have seen that data which argue for the direct
involvement of a cytochrome P-450 in metabolism of
herbicides in the plant is available in only a few cases.
The reasons for this may be many, but studies with
purified plant cytochromes P-450 are reaching a stage
where the enzymology of plant xenobiotic monooxygenases
may be examined in more detail. At the same time,
bacterial monooxygenase systems which mimic the reactions
occurring in higher plants are successfully being used as
model systems to study the biochemistry and biophysics of
herbicide degradation. The cytochromes P-450 of $\underline{S}$.
griseolus provide a system with broad specificity for
sulfonylurea herbicides, and raise a number of mechanistic
questions about cytochromes P-450 in general. The
inducibility of this system suggests that it may be an
excellent model in which to study the biochemical and
genetic steps leading to the expression of cytochromes
P-450 in higher plants that detoxify herbicides. The
information gained in these studies is also valuable in
evaluating the environmental fate of herbicides, since it
provides an understanding of the capacity of soil microorganisms to degrade these compounds.

ACKNOWLEDGMENTS

The authors gratefully acknowledge the scientific collaboration of Antinez V. Jones and Brian A. Wrenn, the technical assistance of Alan L. Stearrett, Wayne L. Alphin, and Roslyn M. Young, and useful discussions with Elmo M. Beyer, Charles J. Arntzen, and Barry L. Marrs. We also thank Professor I.C. Gunsalus for providing a sample of purified cytochrome P-450$_{CAM}$. Contribution number 4158 from the Central Research and Development Department, Du Pont Experimental Station.

REFERENCES

1. COLE, D. 1983. Oxidation of xenobiotics in plants. In Progress in Pesticide Biochemistry and Toxicology. (D.H. Hutson, T.R. Roberts, eds.), John Wiley and Sons Ltd., New York, Vol. 3, pp. 199-254.
2. JOHNSON, I.M., H.W. TALBOT. 1983. Detoxification of pesticides by microbial enzymes. Experientia 39: 1236-1246.
3. JOSHI, M.M., H.M. BROWN, J.A. ROMESSER. 1985. Degradation of chlorsulfuron by soil microorganisms. Weed Sci. 33: 888-893.
4. WESSELS, J.S.C., R. VANDERVEEN. 1956. The action of some derivatives of phenyl urethan and of 3-phenyl-11-dimethylurea on the hill reaction. Biochim. Biophys. Acta. 19: 548-549.
5. STEINRUCKEN, H.C., N. AMRHEIN. 1980. The herbicide glyphosate is a potent inhibitor of 5-enolpyruvyl-shikimic acid-3-phosphate synthase. Biochem. Biophys. Res. Commun. 94: 1207-1212.
6. CHALEFF, R.S., C.J. MAUVAIS. 1984. Acetolactase synthase is the site of action of two sulfonylurea herbicides in higher plants. Science (Wash., D.C.) 224: 1443-1445.
7. LAROSSA, R.A., J.V. SCHLOSS. 1984. The sulfonylurea herbicide sulfometuron methyl is an extremely potent and selective inhibitor of acetolactate synthase in Salmonella typhimurium. J. Biol. Chem. 259: 8433-8457.
8. CHALEFF, R.S., T.B. RAY. 1984. Herbicide resistant mutants from tobacco cell culture. Science (Wash., D.C.) 223: 1148-1151.

9. RYAN, G.E. 1970. Resistance of common groundsel to simazine and atrazine. Weed Sci. 18: 614-616.

10. PFISTER, K., S.R. RADOSEVICH, C.J. ARNTZEN. 1979. Modification of herbicide binding to photosystem II in two biotypes of Senecio vulgaris L. Plant Physiol. 64: 995-999.

11. PFISTER, K., C.J. ARNTZEN. 1979. The mode of action of photosystem II - specific inhibitors in herbicide resistant weed biotypes. Z. Naturforsch. 34c: 996-1009.

12. COMAI, L., L.C. SEN, D.M. STALKER. 1983. An altered aroA gene product confers resistance to the herbicide glyphosate. Science (Wash., D.C.) 221: 370-371.

13. SWEETSER, P.B., G.S. SCHOW, J.M. HUTCHINSON. 1982. Metabolism of chlorsulfuron by plants. Pestic. Biochem. Physiol. 17: 18-23.

14. SWEETSER, P.B. 1985. Safening of sulfonylurea herbicides to cereal crops: mode of herbicide antidote action. Proc. 1985 British Crop Protection Conf.: 1147-1154.

15. ERBES, D.L. 1987. Metabolism of chlorsulfuron in plants: the enzymes involved and the effect of metabolites on acetolactate synthase. Pestic. Biochem. Physiol., in press.

16. ROMESSER, J.A., D.P. O'KEEFE. 1986. Induction of cytochrome P-450 dependent sulfonylurea metabolism in Streptomyces griseolus. Biochem. Biophys. Res. Commun. 140: 650-659.

17. FREAR, D.S., H.R. SWANSON, E.R. MONSAGER. 1983. Acifluorfen metabolism in soybean: diphenylether bond cleavage and the formation of homoglutathione cysteine, and glucose conjugates. Pestic. Biochem. Physiol. 20: 299-310.

18. FREAR, D.S., H.R. SWANSON, F.S. TANAKA. 1969. N-Demethylation of substituted 3-(phenyl)-1-methylureas: isolation and characterization of a microsomal mixed function oxidase from cotton. Phytochemistry 8: 2157-2169.

19. FREAR, D.S., H.R. SWANSON, F.S. TANAKA. 1971. Herbicide metabolism in plants. In Recent Advances in Phytochemistry. (V.C. Runeckles, T.C. Tso, eds.) Academic Press, London, New York, Vol. 5, pp. 225-246.

20. DAWSON, J.H., L.A. ANDERSON, M. SONO. 1983. The diverse spectroscopic properties of ferrous cyto-

chrome P-450-CAM ligand complexes. J. Biol. Chem.
258: 13637-13645.

21. ESTABROOK, R.W., J. PETERSON, J. BARON, A.
    HILDEBRANDT. 1972. The spectrophotometric
    measurement of turbid suspensions of cytochromes
    associated with drug metabolism. In Methods in
    Pharmacology. (D.F. Chignell, eds.), Appleton-
    Century-Crofts, New York, pp. 303-350.

22. PETERSON, J.A., R.A. PROUGH. 1986. Cytochrome P-450
    reductase and cytochrome b_5 in cytochrome P-450
    catalysis. In Cytochrome P-450, Structure,
    Mechanism and Biochemistry. (P.R. Ortiz de
    Montellano, ed.), Plenum Press, New York, London,
    pp. 89-117.

23. GUNSALUS, I.C., J.R. MEEKS, J.D. LIPSCOMB, P. DEBRUNNER,
    E. MUNCK. 1974. Bacterial monooxygenases - The
    P-450 cytochrome system. In Molecular Mechanisms
    of oxygen activation. (O. Hayaishi, ed.), Academic
    Press, New York, London, pp. 559-613.

24. SLIGAR, S.G., R.I. MURRAY. 1986. Cytochrome P-450$_{cam}$
    and other bacterial P-450 enzymes. In P.R. Ortiz
    de Montellano, ed., op. cit. Reference 22, pp.
    161-216.

25. HAMILTON, G.A. 1974. Chemical models and mechanisms
    for oxygenases. In O. Hayaishi, ed., op. cit.
    Reference 23, p. 405.

26. ULLRICH, V., H.J. AHR, L. CASTLE, H. KUTHEN, W.
    NASTAINCZYK, H.H. RUF. 1981. Cytochrome P-450
    as a reductase and oxene transferase: which is
    its characteristic function? In The Biological
    Chemistry of Iron. (H.B. Dunford, D. Dolphin,
    K.N. Raymond, L. Sieker, eds.), I. Reidel Publishing
    Co., Dordrecht, Boston, London, pp. 413-426.

27. COON, M.J., R.E. WHITE. 1980. Cytochrome P-450, a
    versatile catalyst in monooxygenation reactions.
    In Metal Ion Activation of Dioxygen. (T.G. Spiro,
    ed.), John Wiley and Sons, New York, pp. 73-124.

28. JEFCOATE, C.R. 1978. Measurement of substrate and
    inhibitor binding to microsomal cytochrome P-450
    by optical-difference spectroscopy. In Methods in
    Enzymology. (S. Fleischer, L. Packer, eds.),
    Academic Press, New York, San Francisco, London,
    pp. 258-279.

29. BENVENISTE, I., J.P. SALAUN, F. DURST. 1977. Wounding
    induced cinnamic acid hydroxylase in Jerusalem
    artichoke tuber. Phytochemistry 16: 69-73.

30. SALAUN, J.P., I. BENVENISTE, D. REIDHART, F. DURST.
1981. Induction and specificity of a (cytochrome
P-450) dependent laurate in-chain-hydroxylase from
higher plant microsomes. Eur. J. Biochem. 119:
651-655.

31. YOUNG, O., H. BEEVERS. 1976. Mixed function oxidases
from germinating castor bean endosperm. Phytochemistry 15: 379-385.

32. DOHN, D.R., R.L. KRIEGER. 1984. N-Demethylation of
p-chloro-N-methylamiline catalyzed by subcellular
fractions from the avocado pear (Persea americana).
Arch. Biochem. Biophys. 231: 416-423.

33. EARL, J.W., I.R. KENNEDY. 1975. Aldrin epoxidase
from pear roots. Phytochemistry 14: 1507-1512.

34. SAUNDERS, J.A., E.E. CONN, D.H. LIN, M. SHIMADA.
1977. Localization of cinnamic acid 4-monooxy-
genase and the membrane bound enzyme system for
dhurrin biosynthesis in sorghum seedlings. Plant
Physiol. 60: 629-634.

35. HASSON, E.P., C.A. WEST. 1976. Properties of the
system for mixed function oxidation of kaurene and
kaurene derivatives in microsomes of the immature
seed of Marah macrocarpus. Plant Physiol. 58:
479-484.

36. GRAND, C. 1984. Ferulic acid 5-hydroxylase: a new
cytochrome P-450 dependent enzyme from higher
plant microsomes involved in lignin synthesis.
FEBS Lett. 169: 7-11.

37. RICH, P.R., D.S. BENDALL. 1975. Cytochrome components of higher plants. Eur. J. Biochem. 55:
333-341.

38. RICH, P.R., C.J. LAMB. 1977. Biophysical and
enzymological studies upon the interaction of
trans-cinnamic acid with higher plant microsomal
cytochromes P-450. Eur. J. Biochem. 72: 353-360.

39. PFANDLER, R., D. SCHEEL, H. SANDEMANN, H. GRISEBACH.
1977. Stereospecificity of plant microsomal
cinnamic acid 4-hydroxylase. Arch. Biochem.
Biophys. 178: 315-316.

40. GUENGERICH, F.P., G.A. DONNAN, S.T. WRIGHT, M.V. MARTIN,
L.S. KAMINSKY. 1982. Purification and characterization of liver microsomal cytochromes P-450:
electrophoretic, spectral, catalytic, and
immunochemical properties and inducibility of
eight isozymes isolated from rats treated with

phenobarbital or b-naphthoflavone. Biochemistry
21: 6019-6030.

41. BLACK, S.D., M.J. COON. 1986. Comparative structures
of P-450 cytochromes. In P.R. Ortize de Montellano,
ed., op. cit. Reference 22, pp. 161-216.

42. HIGASHI, K., K. IKEUCHI, Y. KARASAKI, M. OBARA. 1983.
Isolation of immunochemically distinct form of
cytochrome P-450 from microsomes of tulip bulbs.
Biochem. Biophys. Res. Commun. 115: 46-52.

43. HIGASHI, K., K. IKEUCHI, M. OBARA, Y. KARASAKI,
H. HIRANO, S. GOTOH, Y. KOGA. 1985. Purification
of a single form of microsomal cytochrome P-450
from tulip bulbs (Tulipa gesneriana L.). Agric.
Biol. Chem. 49: 2399-2405.

44. GABRIAC, B., I. BENVENISTE, F. DURST. 1985. Isolation
and characterization of cytochrome P-450 from
higher plants (Helianthus tuberosis). Comptes
rendus 301, Serie III: 753-758.

45. REICHHART, D., J.P. SALAUN, I. BENVENISTE, F. DURST.
1979. Induction by manganese, ethanol, phenobarbital, and herbicides of microsomal cytochrome
P-450 in higher plant tissues. Arch. Biochem.
Biophys. 196: 301-303.

46. REICHHART, D., J.P. SALAUN, I. BENVENISTE, F. DURST.
1980. Time course of induction of cytochrome
P-450, NADPH-cytochrome c reductase and cinnamic
acid hydroxylase by phenobarbital, ethanol,
herbicides and manganese in higher plant microsomes.
Plant Physiol. 66: 600-604.

47. KOTAKE, A.N., Y. FUNAE. 1980. High performance
liquid chromatography technique for resolving
multiple forms of hepatic membrane-bound cytochrome
P-450. Proc. Natl. Acad. Sci. USA 77: 6473-6475.

48. BANSAL, S.K., J.H. LOVE, H.L. GURTOO. 1984. Resolution of multiple forms of cytochrome P-450 by high
performance liquid chromatography. J. Chromatogr.
297: 119-127.

49. FUNAE, Y., R. SEO, S. IMAOKA. 1986. Two step purification of cytochrome P-450 from rat liver microsomes
using high performance liquid chromatography.
J. Chromatogr. 374: 271-278.

50. THOMAS, P.E., D. RYAN, W. LEVIN. 1976. An improved
staining procedure for the detected of the
peroxidase activity of cytochrome P-450 on sodium
dodecyl sulfate polyarrylamide gels. Anal. Biochem.
75: 168-176.

51.   THOMAS, P.E., J. REIDY, L.M. REIK, D.E. RYAN, D.R.
      KOOP, W. LEVIN. 1984. Use of monoclonal antibody
      probes against rat hepatic cytochromes P-450c and
      P-450d to detect immunochemically related
      isozymes in liver microsomes from different
      species. Arch. Biochem. Biophys. 235: 239-253.
52.   HENDRY, G. 1986. Why do plants have cytochrome
      P-450? Detoxidation versus defense. New Phytol.
      102: 239-247.

Chapter Seven

PCBs IN THE ATMOSPHERE AND THEIR ACCUMULATION IN FOLIAGE
AND CROPS

EDWARD H. BUCKLEY

Boyce Thompson Institute for Plant Research
Cornell University
Ithaca, New York 14853

INTRODUCTION

During the late 1950's and 1960's, when gas chromatographic separations and analyses became important analytical
tools, a major problem was the presence of numerous unknown
compounds that interfered in the research and monitoring of
pesticides. In 1966, Jensen[1,2] showed that in the Baltic
Sea, many of these interfering compounds were polychlorinated biphenyls (PCBs) - a whole family of industrial
compounds with physical properties similar to DDT and DDE.
Although PCBs had never intentionally been introduced into
the environment they had become common as had pesticides.

Sources of airborne PCBs are numerous and diverse,
since PCBs were used as dielectric fluids in capacitors

and transformers since 1930, and as stable industrial
fluids in hydraulic equipment, turbines, vacuum pumps and
in high temperature heat transfer systems. After the
second world war, PCBs were also used extensively as
plasticizers and spreaders for industrial and home use
in surface coatings, plastics, waxes, adhesives, sealants,
printing ink, copy paper and in dyes for textiles,
upholstery and wallpaper, and in some cases, as a fire
retardant in fabrics.

There have been numerous reviews of polychlorinated
biphenyls over the past fifteen years or so. Two major
compilations of PCB information are The Chemistry of PCBs
by Hutzinger, Safe and Zitko,[3] and the more recent 1979
report by the National Research Council, Polychlorinated
Biphenyls.[4] However, both of these publications contain
little information on PCBs in terrestrial plants. A most
informative review of PCBs in rooted plants is that by
Strek and Weber.[5]

The subject of this review — PCBs in the atmosphere
and their accumulation in foliage and crops — is only a
small part of the research on PCBs in plants. Inadver-
tently, and initially without man's knowledge, a global
experiment with PCBs has been performed. The purpose of
this review is to integrate some of the information on PCB
accumulation in foliage and crops from that experiment
with the larger literature in this field.

We can look at PCBs as a family of model compounds
that are readily detected, but have physical properties
similar to many other anthropogenic compounds that are
much more difficult to detect in the environment, especially
in plants. Cyclic aromatic hydrocarbons, for example, are
numerous and diverse in plants, and can be extremely
difficult to separate and quantitate.

There are two significant observations to be
presented here. The first is that compounds used in
manufacturing products for man's convenience, comfort,
and aesthetic tastes can also turn up in his food.
Secondly, as phytochemists we must be aware that there may
be numerous compounds in a plant that do not arise from the
plant's biosynthetic capacity, but instead, are of anthropogenic origin or had a precursor of anthropogenic origin.

PCBs IN AIR

The low volatility of PCBs is well documented;[3] nevertheless, airborne transport and atmospheric fallout of PCBs in rain, snow, and particulates is now known to be a significant source of PCBs in the Great Lakes.[6,7] Furthermore, airborne transport of PCBs from their various sites of use and of disposal is considered to be a primary mode for their global distribution.[4] Even the snow of both polar caps contains PCBs. One factor that helps to explain this apparent paradox of extensive airborne transport of compounds of relatively low volatility is that PCBs volatilize more readily from a film of water than their vapor pressures would indicate.[8] PCBs are hydrophobic substances with very low solubility in water, but also with very low retention in water. Thus, a film of water can act as a carrier that enhances evaporation of PCBs, whereas a layer of water can depress their evaporation. However, if a layer of water contains suspended particulates (especially organic particulates and colloids) that are contaminated with PCBs, then evaporation of PCBs is enhanced because the particulates are in dynamic equilibrium with the dissolved PCB phases. Consequently, the former dissolved PCBs that evaporated into the atmosphere are rapidly replaced by more dissolved PCBs that also evaporate. When PCB levels in the atmosphere are proportionately greater than those in water, suspended particulates in water greatly enhance the ability of that body of water to trap vapor-phase PCBs and PCBs from precipitation. Hence, the long-term fate for PCBs in the environment is for them to be attached to a particulate that sinks to the bottom of an ocean.

Reports on the atmospheric transport of PCBs over the North Atlantic, the Gulf of Mexico, and the North Pacific Ocean, indicate that these compounds remain airborne over very long distances.[9-12] Although they predominate in the northern hemisphere, they appear to be present throughout the global atmosphere.[4]

Airborne PCBs exist both in the vapor-phase and attached to particulates in the atmosphere. The usual technology for measuring airborne PCBs is to collect the particulates from the sampled air stream on a glass fiber filter. However, no one has devised a means of determining how many PCBs are associated with particulates that are too

small to be collected on the filters, or how to measure
the degree to which PCBs in the vapor-phase are adsorbed
onto the collected particulates while on the filter, or
are volatilized off the collected particulates. These
problems have been variously addressed,[4] and there is an
element of consensus that more than 99% of the atmospheric
PCBs over rural continental areas and over oceans are in
the vapor-phase. Concentrations there range from 0.002
ng/m^3 (which is extremely difficult to measure) to 1.6
ng/m^3. Airborne PCBs are appreciably more abundant over
urban/metropolitan areas, in the general range of 0.5
ng/m^3 to 40 ng/m^3, with an appreciable component (up to
50%) being attached to particulates. Interestingly, our
homes often contain airborne PCBs (primarily in the
vapor-phase) at concentrations of several hundred
ng/m^3.[13] When we open windows to let in the fresh air,
we pollute the atmosphere - and our backyard gardens,
as will be mentioned later.

THE DIVERSITY IN PCBs

    Basic information on commercial PCB preparations,
their physical properties and commercial synthesis,
synthesis of individual PCB congeners, chemical reactions,
photodegradation, and procedures for analyzing PCBs have
been compiled in The Chemistry of PCBs.[3] Several
characteristics of PCBs are particularly relevant to
this discussion of airborne PCBs that accumulate in
vegetation, and these are:

    1. PCBs are a family of compounds which contain the
biphenyl ring system as their basic carbon skeleton.
There are 10 possible sites for chlorination on the
biphenyl ring system; therefore there are 10 groups of
PCB isomers, from the three monochlorobiphenyls to the
one decachlorobiphenyl. There are 46 possible penta-
chlorobiphenyl isomers.[4] Collectively, the 10 groups
of PCB isomers are called PCB congeners. Of the 209
possible PCB congeners, there are approximately 70 known
congeners in the various commercial preparations and in
the environment, although some of the lesser chlorinated
congeners are disappearing. (Note: electron-capture
detectors which are usually used to measure PCB congeners
during GLC are two orders of magnitude less sensitive in
measuring monochloro congeners than trichloro and tetra-

chloro congeners. This contributes appreciably to the
"apparent" disappearance of the lesser chlorinated congeners.)

   2. The Monsanto Company synthesized almost all of
the commercial mixtures of PCBs used in North America,
and marketed them under the trademark Aroclor, followed
by four digits. The first two digits for PCBs were 12,
to designate the 12 carbons of the biphenyl ring. The
second two digits represented the percent chlorination.
Each commercial preparation contained five or six isomer
classes which gave each mixture a wide range of physical
properties. The composition of the different mixtures
can be described approximately: Aroclor 1221, primarily
mono- and dichloro congeners; Aroclor 1232, primarily di-
and trichloro congeners; Aroclor 1242, primarily tri-
and tetrachloro congeners; Aroclor 1248, primarily tetra-
and pentachloro congeners; Aroclor 1254, primarily
penta- and hexachloro congeners; and Aroclor 1260,
primarily hexa- and heptachloro congeners.

   3. Solubilities in water of the individual congeners
vary by three orders of magnitude from 5.9 mg/l (ppm) for
2-monochlorobiphenyl to 0.0070 mg/l for 2,2',3,3',4,4',5,5'
octachlorobiphenyl (0.015 mg/l for decachlorobiphenyl).[4]
Water solubility tends to decrease as the degree of
chlorination increases, but the steric position of the
chlorines also greatly influences water solubility. Also,
the reported values for solubility vary appreciably,
particularly for those congeners with very low solubility
where measurement is extremely difficult. In this
connection, it is relevant to point out that Gschwend
and Wu[14] do not believe that it is yet possible to
measure the water solubility of a highly hydrophobic
compound.

   4. Evaporation of PCBs from a source is influenced
greatly by several factors: the congeners involved, the
presence of organic matter, moisture, the temperature and
air movement (which involves the depth and the porosity
of the PCB source). In general, the less chlorinated
congeners are the more volatile. Organic matter tends
to bind PCBs, particularly the more highly chlorinated.
As mentioned previously, films of moisture enhance
volatility. Temperature generally increases volatility,
but can decrease it by drying the substrate. Air movement
prevents a state of equilibrium. Therefore, a fine-grained,

low-porosity substrate allows a vapor-phase equilibrium
to be approached, while a coarse, porous substrate
enhances volatility, as does a surface wind. Consequently,
the relative abundance of congeners in the vapor phase
leaving a source is quite different from the relative
quantities of those same congeners at the source.
However, under many diverse conditions it is generally
believed[5] that the amount of lower chlorinated congeners
in the vapor phase is always much greater than in the
adsorbed state at the source.

ACCUMULATION OF VAPOR-PHASE PCBs IN PLANTS

In the late 1950's, Lichtenstein[15] proposed that
pesticides, such as DDT, that persisted in the soil could
contaminate future crops by vaporizing from the soil and
being absorbed by the foliage even though the acropetal
translocation from roots to above-ground structures was
relatively small. This problem was studied during the
1960's[16,17] and data were acquired that is applicable to
PCBs.

Klein and Weisgerber[18] published the first information
that suggested that terrestrial plants accumulate airborne
PCBs. Since then, both mosses[19-22] and lichens[23,24] have
been used to monitor airborne PCBs. A recent paper
describes the use of both soil and grass to monitor PCBs
in rural and urban areas of the United Kingdom.[25] The data
of Klein and Weisgerber[18] also suggested that the amount
of PCBs accumulated in vegetation was species dependent
as well as concentration dependent, a typical situation
found in past experience with other airborne pollutants.
Consequently, when we collected foliar samples in the
field, locations were selected where two or more species
could be sampled in the same space (usually within the
same cubic meter of space) so that if they existed,
ratios of PCB accumulation among species could be
determined.

Background Level (B Values)

Background levels of PCBs in leaves (leaflets of
compound leaves) and in hay crops were defined in areas
where we consistently found the lowest levels of PCBs
in vegetation. These areas were located in Washington

Table 1.   Background concentrations of PCBs in vegetation
in Washington and Saratoga Counties, New York, September,
1979.

Species	Sample[1]	Total PCBs (ppm)[2]
Pinus strobus L. (white pine)	leaves (9)	0.026 ± 0.005
Medicago sativa L. (alfalfa)	hay (12)	0.045 ± 0.015
Trifolium pratense L. (red clover)	hay (6)	0.050 ± 0.013
Fraxinus americana L. (white ash)	leaflets (6)	0.054 ± 0.010
Pinus rigida Mill (pitch pine)	leaves (6)	0.054 ± 0.011
Acer rubrum L. (red maple)	leaves (6)	0.061 ± 0.009
Lolium perenne L. (perennial rye)	hay (6)	0.075 ± 0.012
Zea mays L. (corn)	leaf blade above upper ear (18)	0.077 ± 0.006
Zea mays L. (corn)	silage (6)	0.022 ± 0.003
Zea mays L. (corn)	grain on ears (6)	<0.001
Populus tremuloides Michx. (trembling aspen)	leaves (8)	0.088 ± 0.013
Populus grandidentata Michx. (large-toothed aspen)	leaves (6)	0.089 ± 0.008
Quercus rubra L. (red oak)	leaves (6)	0.089 ± 0.015
Phleum pratense L. (timothy)	hay (8)	0.091 ± 0.014
Rhus typhina L. (staghorn sumac)	leaflets (14)	0.096 ± 0.007
Rhus glabra L. (smooth sumac)	leaflets (6)	0.104 ± 0.008
Bromus inermis Leyss. (brome grass)	hay (6)	0.117 ± 0.022
Dactylis glomerata L. (orchard grass)	hay (6)	0.119 ± 0.026
Solidago graminifolia L. (goldenrod (gram.))	leaves (15)	0.250 ± 0.045
Solidago nemoralis Ait. (goldenrod (nem.))	leaves (6)	0.288 ± 0.027

[1]Number of samples is shown in brackets.

[2]ppm per unit dry weight of sample.

and Saratoga Counties, New York, in 1978 and 1979, and
were found to be comparable to more than twenty other
background areas sampled across New York State in 1980.
Background PCB concentrations in vegetation for September
1979 are shown in Table 1 for 17 species. The analyses of
hay crops are more variable than the analyses of leaves,
due to the variable growth structure and degree of heading
in the hay crops.

   That background levels of PCBs were dropping rapidly
in New York State in 1978, 1979, and 1980 is shown in

Table 2A. Decrease in background concentrations of PCBs in goldenrod <u>Solidago graminifolia</u> in eastern and central counties of New York State.

Location	Sept. 1978	ppm (dry weight) Sept. 1979	Sept. 1980
Eastern NY[1] (Washington & Saratoga Co.)	(4) 0.323 ± 0.026	(4) 0.258 ± 0.035	(4) 0.178 ± 0.025
		(5) 0.250 ± 0.057	(5) 0.188 ± 0.024
		(6) 0.245 ± 0.049	
Mean of total		(15) 0.250 ± 0.045	(9) 0.183 ± 0.023
Central NY (Tompkins Co.)	(2) 0.32 ± 0.04	(4) 0.255 ± 0.034	(4) 0.170 ± 0.029

[1]Number of samples and sites is shown in brackets.

Table 2B. Decrease in background concentrations of PCBs in trembling aspen in eastern and central New York State.

Location	Sept. 1978	ppm (dry weight) Sept. 1979	Sept. 1980
Eastern NY[1] (Washington & Saratoga Co.)	(2) 0.12 ± 0.01	(2) 0.09	(2) 0.07 ± 0.01
		(6) 0.087 ± 0.015	(6) 0.063 ± 0.015
Mean of total		(8) 0.088 ± 0.013	(8) 0.065 ± 0.014
Central NY (Tompkins Co.)	(4) 0.108 ± 0.010	(4) 0.088 ± 0.010	(4) 0.058 ± 0.010

[1]Number of sample and sites is shown in brackets.

Tables 2A and 2B using leaves of goldenrod (<u>Solidago graminifolia</u> L.) and leaves of trembling aspen (<u>Populus tremuloides</u> Michx.) from Washington and Saratoga Counties in eastern New York and Tompkins County in central New York.

Background levels of PCBs were measured in leaves of field corn collected from twenty-five dairy areas across

New York State in September, 1980.[26]   They had concentrations ranging from 0.03 to 0.06 ppm total PCBs, with a
mean of 0.044 ± 0.010, which indicated appreciable
uniformity across New York State in 1980. These data
for New York State, combined with data from Tables 2A
and 2B, suggest that background levels across New York
may have been appreciably higher in the 1970's. The
evidence for falling background levels (Tables 2A and
2B) was welcomed by Bidleman[9] who was still involved with
monitoring the atmosphere over the Atlantic Ocean in the
late 1970's. He had expected to see an appreciable
decrease in airborne PCBs in the early 1970's when many
PCB sources were eliminated, but that decrease had not
started until the late 1970's. He had observed a rapid
drop in the averaged air concentrations, but the variability of the atmospheric data was so great that a
statistically significant difference had not been
reached at that time. Villeneuve[24] also observed no
apparent reduction in airborne PCBs in the early 1970's.
He reported twelve years of increasing PCB accumulation
in lichens in Sweden from 1961 through 1972. He found
no decrease in the accumulation of PCBs in lichens, even
though the use and the apparent major sources of PCBs
were substantially reduced in Sweden in 1970. Therefore,
there is some confidence in the observations that the
decrease in the use of PCBs in North America during the
early 1970's did not result in decreases in airborne PCBs
until the late 1970's, and that the rapid change in
background levels of PCBs in vegetation that was observed
in New York State from 1978 to 1980 was related to a real
decrease in background levels of atmospheric PCBs at that
time.

Multiples of Background Level (MBL Values)

For the purpose of presenting information, background
levels of PCBs in vegetation were discussed before
multiples of background level, but they evolved in the
opposite order. Our studies of foliar uptake of PCBs were
initiated to determine if vapor-phase PCBs that evaporated
from known PCB sources were contaminating adjacent vegetation. As expected, contamination in vegetation near
some of the sites was quite high, but decreased quickly in
a manner that allowed curve fitting to the simple formula
below, developed specifically for estimating long-term

(seasonal) average concentrations of vapor-phase pollutants from a point source.[27]

$$\overline{X}_A = \overline{X}_{AC} \left(\frac{x_c}{x} \right)^{p + 1}$$ Equation 1

$\overline{X}_A$ is the long-term average concentration of a vapor-phase air pollutant at a point x units from the source; X_{AC} is the known long-term average concentration of the vapor-phase air pollutant at a long-term monitoring point x_c units from the source; p is an atmospheric dispersion constant that must be determined to fit the long-term average atmospheric dispersion conditions along that particular transect from the source. Generally, dispersion factors range from 0.1 (low dispersion) to 0.5 (high dispersion).

When the foliar PCB data, expressed in ppm, was entered into the equation, the fit was highly erratic unless the data were limited to a single species. However, a range of species could be used in the equation if allowances were made for the differences in foliar PCB accumulation among species. We found that the differences among plant species in their foliar PCB accumulation (i.e., the ratios of accumulation between species) were the same, whether at high levels of contamination (fumigation from a vapor-phase PCB source) or at background level. Consequently, by simply dividing the foliar PCB content found near a PCB source by the foliar PCB content for that species at background level, we had data expressed in multiples of background level (MBL values), and these MBL values were comparable among species that shared the same space.[28] In all of these cases, the soil occupied by the roots of the plants was not contaminated with PCBs. Surface litter over the soil did contain low levels of PCBs throughout the growing season. PCBs in crop soils were negligible in the areas studied.

The use of MBL values for foliar PCB concentration made it possible to calculate dispersion constants along various transects from a known PCB source. In this manner, secondary PCB sources were often detected. Secondary sources frequently encountered were roads that had been

oiled to reduce dust, or roads where dredge spoils from
the Hudson River had been used for sanding in the winter.

Originally we ran a few transects for miles, trying
to find the low levels predicted by equation 1, and that
was how we found a background level. Subsequently we
showed that the high background levels in Washington and
Saratoga Counties were comparable to background levels
across the State of New York,[26] the northeast and Michigan
State.

MBL values as high as 700 have been shown to retain
their validity.[28] Now that background levels have
decreased, while the proportionality among species has
not changed, the former MBL values[28] could be recalculated
(same concentration in ppm ÷ new lower background values).
Thus, with the same data base for polluted sites combined
with new background values, the updated MBL values could
appreciably exceed 1000 and still retain their validity
(i.e., same ppm limit but higher MBL value).

MBL values can also be used to measure vertical
gradients. Trees growing in uncontaminated soil, or
alternatively a community of plants that includes trees,
will show an appreciably vertical gradient if there is a
nearby ground-level source of PCB. Proximity to a ground-
level source can be determined readily using vertical
foliar PCB concentrations (or MBL values, when more than
one species is involved). In practice, a sample at 1-
meter height and another at 10-meter height would discern
at least a 25% differential at a distance of 70 meters
from a ground-level source, and that differential would
increase rapidly closer to the source. We have had no
experience with PCB sources that are physically located
higher than ground level. (Our PCB sources did not
include tall smoke stacks which are much more complex
sources to evaluate and predict.)

There are practical uses for the MBL system other
than for monitoring or detecting PCB sources. An MBL
system developed for crops would provide guidance to
growers operating near sources of vapor-phase PCBs. For
example, crops of corn silage accumulate lower quantities
of vapor-phase PCBs than hay crops (Table 1), while crops
of corn grain or corn on the ear accumulate only a very
small amount of PCBs. The FDA limit for PCBs in cattle

feed is 0.2 ppm.[29] As will be shown later, this level of
PCB contamination is reached at surprisingly low concentrations of PCBs in air, and in special situations can be
a serious concern to dairy farmers. The relationship
between PCBs in cattle feed and PCBs in milk fat (FDA limit
1.5 ppm PCBs) has been documented,[30-32] and shows a four-
to five-fold concentration in the milk fat for Aroclor 1254
which contains congeners that degrade very little in the
rumen of the cow.

METABOLISM OF PCBs IN PLANTS

   Metabolism of PCBs in plants,[33-38] and in lichens[39]
has been studied and reviewed.[5,40] In summary, the
metabolism of PCBs in plants is very slow, and essentially
limited to mono-, di-, and trichloro isomers. The higher
chlorinated congeners are nearly inert. The metabolism
that has been detected in plants is hydroxylation of one
or both rings, and then adducts are formed. Sugar adducts
of hydroxylated PCBs may be a factor in the translocation
of low-chlorinated congeners; however, it has been difficult to define the adducts. When ^{14}C-labeled PCBs are
added in solution to leaves, more than 90% evaporates
(disappears), while translocation is low and metabolism
appears negligible.[35,41] The evaporation of ^{14}C-PCBs
applied to the leaves is supportive of our research
findings (unpublished) that foliage does not directly
accumulate additional PCBs from rain.

   The fact that PCBs are slowly metabolized in plants
has been a factor in their use as PCB monitors in aquatic
ecosystems.[42,43] Plants accumulate PCBs from their
environment, but do not modify their extractable congener
profiles appreciably. Of course, the accumulation process
is selective, and that bias has to be recognized if
quantitative accuracy is required.

   More recently, the uptake by plant tissue cultures of
a mono- and a dichlorobiphenyl was studied and the uptake
correlated with the lipid content of the cultures.[44] More
significantly, the tissue cultures showed responses ranging
from highly toxic to negligible effects, which were not
correlated with PCB uptake. This suggests that tissue
cultures may be a rewarding way to study plant metabolism
of PCBs. Such experiments would be free from microbial

populations that are so significant in both the soil and
aquatic environments used in previous PCB metabolic studies
with plants.

EFFECTS OF PCBs ON PLANT GROWTH

Aquatic plants, particularly algae, appear much more
sensitive than terrestrial plants to the presence of PCBs.
However, there is growing evidence that the wide variability
among algal species, in terms of their inhibition by PCBs
of basic photosynthetic, reproductive, and growth processes,
may have some counterpart among terrestrial species.

A series of studies have been made of the effects
of PCBs upon algal communities[45-53] at "real world" PCB
concentrations (ppb or less in water), demonstrating the
great range of sensitivity among algal species. As
mentioned previously, plant tissue cultures, suspended in
liquid media, also showed a great range of responses.[44]
At higher concentrations, algal photosynthesis is depressed
and chloroplast membranes and lamellae are distorted.[54,55]
By contrast, PCBs are relatively non-toxic to Chlorella,
although they are concentrated by the cell.[56] When the
aquatic plant Spirodella oligorrhiza (Kurtz) Hegelm. was
exposed to high aquatic concentration of PCBs, it
became chlorotic. The chloroplasts were enlarged, the
thylakoids and stroma became irregular, starch grains
enlarged and plastoglobuli formed within them, while the
other components of the cells appeared structurally
intact.[57]

The chloroplasts of higher plants are assumed to be
susceptible to PCBs, based upon the proven susceptibility
of photosynthetic systems in many algae. However, the
only evidence supporting this hypothesis is work with
isolated spinach chloroplasts.[58] Oxygen evolution was
inhibited by Aroclor 1221 (low chlorinated PCB congeners
like those known to be translocated in plants), although
there was no inhibition of photosystem I when it was
examined independently. Therefore, inhibition at a site
on the electron transport chain close to photosystem II
was implicated.

Generally, terrestrial vegetation has been considered
a carrier of PCBs that is not affected by them. However,

that concept is changing. Weber and Mrozek[41] were the first
to demonstrate inhibition of growth caused by high concentrations (up to 1000 ppm) of PCBs in soil. Soybeans were
inhibited more than fescue. Inhibition was accompanied
by an increase in transpiration. Weber and Mrozek also
showed a dramatic reversal of these symptoms by mixing
activated carbon with the PCB-contaminated soil. Subsequently, beet plants were also shown to be susceptible,
and pigweed very susceptible, while corn and sorghum were
not affected by soil concentrations as high as 1000 ppm
PCBs.[59,60]

The only evidence of PCBs causing somatic mutations
in plants is in ostrich ferns,[61] although the real causal
agent could be impurities in the PCBs. Surprising is the
well documented report that the growth form of the marsh
plant Spartina alterniflora Loisel is modified with just
1 ppm Aroclor 1254 in the sediments.[62] Finally there is
the unpublished report that the marsh plant Scirpus
lacustris increased its growth in water containing several
hundred ppb PCBs while simultaneously decreasing the PCB
concentration in its aquatic environment.[62a]

The message here is that higher plants are not just
carriers of PCBs, but respond in different ways as a
function of their unique characteristics. At present,
however, growth effects in terrestrial plants are being
noted when the PCB source is in the soil rather than in
the air.

ROOT UPTAKE AND TRANSLOCATION

Strek and Weber reviewed root uptake and translocation of PCBs as previously mentioned. Some aspects of
this subject should be summarized here because of their
relevance in interpreting crop and foliar accumulation
of PCBs.

Under normal conditions, PCBs in soil will contaminate plant roots.[37,63-65] However, the use of activated
carbon can essentially bind all PCBs in the soil to make
them unavailable to the roots,[39,59,60] as well as to curtail
evaporation of PCBs from the soil. Once PCBs are in the
roots, the degree of acropetal translocation may be quite
variable depending upon the species.[59,64-66] In some

cases, the acropetal translocation may be extensive, as in
the marsh plant <u>Spartina</u> <u>alterniflora</u> Loisel[67-69] and
<u>Lythrum</u> <u>salicaria</u> L. In the latter case, 2 mono- and 2,2'-
dichlorobiphenyl were observed to volatilize from the
leaves.[70] In other cases, acropetal translocation is very
restricted.[71-75]

Therefore, unless the inability of the species to
translocate PCBs is known, the use of the MBL system for
monitoring vapor-phase PCBs or for predicting crop contamination should be restricted to sites where the soil is not
contaminated with PCBs.

MBL VALUES AND VAPOR-PHASE PCB CONCENTRATIONS

An approximation was made of the relationship between
MBL values and the mean seasonal vapor-phase concentration
of PCBs in the atmosphere. To do this, a continuous air
monitoring system for vapor-phase PCBs was set up from the
beginning of June to nearly the end of September. The
system consisted of a PCB source (a compact, overgrown,
former PCB dumpsite), and two miles north of it, a background area. Four continuous air monitoring sites were
located in the background area and another four continuous
air monitoring sites were located along a northeasterly
radial transect extending out from the dumpsite. Their
distances from the dump were 62 m, 76 m, 116 m, and 215 m.
Vapor-phase PCBs were monitored according to the method
of Bidleman and Olney[76] which uses polyurethane foam plugs
as an absorbent of PCBs. Three plugs were used in each
sampling column, and they were changed at weekly intervals.
The first plug collected the sample, while the second and
third plugs were blanks except in a few cases during the
hottest weeks when there was some sample breakthrough into
the second plug. Air sampling rates near the PCB source
were regulated at each sampling column by a calibrated
limiting orifice, and measured daily with a flowmeter to
check for evidence of malfunctions in the system. Air
sampling in the background area was at full pump capacity,
and flow rates were measured with calibrated ball flow-
meters at least twice a day, one of the times being at
night or before sunrise. Each sampling pump was metered to
record total hours of operation. Leaves of staghorn sumac
and smooth sumac were collected in September. At the PCB
source site, all foliar samples were collected on the same

transect line as the four air samples sites. Foliar
samples were placed in plastic sample bags, frozen in a
dry ice chest, transported frozen, freeze-dried, milled,
then extracted overnight (approximately 16 hours) with
hexane in Soxhlet extractors. Clean-up of each sample on
a florisil column was performed before gas chromatographic
analysis, using a packed column and electron-capture
detector.

Table 3. Calibration of PCBs in foliage of staghorn sumac
vs. mean seasonal vapor-phase PCBs in air.

Foliar PCBs ppm	Distance m	Measured seasonal air conc., ng/m^3	Calc. seasonal air conc., ng/m^3
air sampler	62	305	298
3.84 ±0.13 (5)	71	-	253
air sampler*	76	233*	233*
3.00 ±0.17 (5)	95	-	178
air sampler	116	142	139
2.12 ±0.13 (5)	124	-	128
1.40 ±0.09 (5)	180	-	81
1.07 ±0.06 (5)	210	-	67
air sampler	215	65.2	65.5
0.92 ±0.06 (5)	230	-	60

0.059 ±0.005 (10) background   0.3   -

*main reference sampler

$y = 64.3x - 4.3$ when y = ng/m^3, x = ppm PCBs in foliage

Number of samples is shown in brackets.

Table 4.   Calibration of PCBs in foliage of smooth sumac
vs. mean seasonal vapor-phase PCBs in air.

Foliar PCBs ppm	Distance m	Measured seasonal air conc., ng/m^3	Calc. seasonal air conc., ng/m^3
5.27 ±0.20 (5)	60	-	311
air sampler	62	305	298
4.28 ±0.13 (5)	71	-	253
air sampler*	76	233*	233*
3.52 ±0.40 (5)	91	-	187
air sampler	116	142	139
2.04 ±0.17 (5)	130	-	121
2.02 ±0.10 (5)	140	-	111
air sampler	215	65.2	65.5
0.066 ±0.005 (10) background		0.3	-

*main reference sampler

$y = 59.0x - 5.0$ when $y = $ ng/m^3, $x = $ ppm PCBs in foliage

Number of samples is shown in brackets.

A dispersion constant of p = 0.22 was used to fit
Equation 1 to the data for staghorn sumac (Table 3) and for
smooth sumac (Table 4).   This distance x_c could have been
any of the four continuous air monitoring sites, but the
site at 76 meters was chosen and the data from that site
are marked with an asterisk.

Note that 10 samples were collected from the background area since the accuracy of that measurement was
vital.   Therefore in the calculation of the linear regression, the mean value for each sampling point near the PCB
source was entered once (five samples per sampling
station) while the mean background level was entered

twice (ten samples). Therefore n = 8 for the staghorn
sumac regression and n = 7 for the smooth sumac regression.

Our most extensive background data by far was obtained
in 1979. To utilize those data, the foliar PCB data (in
ppm) for the two sumac species was converted to 1979 MBL
values establishing a relationship between 1979 MBL values
and the mean of seasonal air concentrations (Tables 5A and
5B). The fact that actual background levels have decreased
significantly since 1979 does not affect the usefulness of
the 1979 values which simply provide a conversion factor
from one species to another.

The close fit of the two data sets in Tables 5A and
5B should not be misconstrued as an indication of their
accuracy. In addition to good luck, all of the sampling
points were along a single transect which theoretically
fumigated all of the plants at the same intervals. Of
course the dispersion factor p changed constantly so that
the relative intensities were different but approximated
a dispersion factor of 0.22 over the growing season. If
ten radial transects had been established around the source
site, each would have had a different frequency and
intensity of fumigation which could influence the results.
Different dispersion factors for each direction would not
be a factor as long as each was essentially uniform over

Table 5A. Relationship between PCB accumulation in
staghorn sumac leaflets in September, MBL units[*], and the
averaged concentration of vapor-phase PCBs June through
September.

Foliar PCBs (ppm)	Staghorn sumac MBL units	Vapor-phase PCBs (ng/m^3)
0.096	1.0	1.9
0.200	2.1	8.1
0.960	10.0	58.
1.92	20.0	119.
2.88	30.0	181.

*MBL units - see Table of Contents

Table 5B. Relationship between PCB accumulation in smooth
sumac leaflets in September, MBL units*, and the averaged
concentration of vapor-phase PCBs June through September.

Foliar PCBs (ppm)	Smooth sumac MBL units	Vapor-phase PCBs (ng/m^3)
0.104	1.0	1.2
0.200	1.9	7.3
0.960	10.0	56.
1.92	20.0	117.
2.88	30.0	179.

*MBL units - see Table of Contents

the entire length of the transect when downwind of the PCB
source.

Controlled releases of vapor-phase PCBs are required
for accurate field work. The fumigation episodes must be
distinct and monitored and studied as distinct episodes.
What I have presented here is only encouraging preliminary
evidence that PCBs are a good model for the study of crop
contamination and the unique characteristics of plant
species.

The applied message from this work is the surprisingly low concentration of vapor-phase PCBs that is
sufficient to contaminate a hay crop beyond the FDA limits
for cattle feed. It is five orders of magnitude below
the airborne PCB standards set for humans by OSHA. It
is an order of magnitude below the levels that are common
in our homes and work places. I wonder how much of the
elevated accumulations of PCBs in vegetation in urban
areas, even in hamlets, is due to PCBs in our buildings?
It would be illegal to feed a cow the fresh leafy vegetables that many of us grow in our home gardens.

SUMMARY

As a matter of perspective, PCBs are a problem in the environment, but perhaps less serious a problem than originally thought. The main thing we can learn from the global PCB experiment is that compounds with similar physical properties will disperse globally and enter the food chain of almost all creatures on earth. We all have a legitimate interest in the lesson that PCBs can teach us about responsible production and use of anthropogenic compounds.

PCBs are ideally suited for study because they are relatively easy to detect, they are dispersed widely in the environment, and the seventy-plus congeners provide wide ranges of physical properties for study. It is not advisable to measure only total PCBs because the analyses are based upon the more highly chlorinated congeners of Aroclor 1242. By the standard technique that Aroclor profiles are "quantitated" to obtain total PCBs, the information reported here is based 80 to 90% on tri- and tetrachloro congeners, with the remainder on pentachloro congeners. Monochloro to heptachloro congeners were present, but their partitioning was much different.

In summary, we have learned that the vapor-phase PCB content in buildings and in our communities exceeds the concentration required to contaminate a hay crop beyond the FDA limit of 0.2 ppm PCBs for cattle feed. We have learned that the PCB levels are decreasing significantly, but will be sufficient for research for another two or three decades. Vapor-phase PCBs in the atmosphere appear to be contaminating the foliage of all terrestrial vegetation, but the degree of foliar contamination varies by at least an order of magnitude from species to species. The contamination of harvestable crops by vapor-phase PCBs is much more variable, certainly exceeding two order of magnitude. There is now increasing evidence that such magnitudes of variation may also be true for above-ground crops contaminated from PCBs in soil. Therefore, we must be more aware that chemical compounds we produce for non-agricultural purposes can become part of our diet. Also, as phytochemists, we must be aware that compounds of anthropogenic origin, not meant for agriculture, are present in plants in measurable quantities.

ACKNOWLEDGMENTS

The technical assistance of J.J. Egnaczak and J.M. Buckley of Boyce Thompson Institute is gratefully acknowledged, as is the interest and counsel of Professor D.J. Lisk and W.H. Gutenmann of the Toxic Chemicals Laboratory, Cornell University. The study on accumulation of airborne PCBs in foliage was supported jointly by the New York State Department of Environmental Conservation, (Contract C-143563) and by the Boyce Thompson Institute.

REFERENCES

1. JENSEN, S. 1966. Report of a new chemical hazard. New Sci. 32: 612.
2. JENSEN, S., A.G. JOHNELS, M. OLSSON, G. OTTERLIND. 1969. PCB in marine animals from Swedish waters. Nature (London) 224: 247-250.
3. HUTZINGER, O., S. SAFE, V. ZITKO. 1974. The Chemistry of PCBs. CRC Press, Cleveland, Ohio,
4. NATIONAL RESEARCH COUNCIL. 1979. Polychlorinated Biphenyls. National Academy of Sciences, Washington, D.C.
5. STREK, H.J., J.B. WEBER. 1982. Behaviour of poly-chlorinated biphenyls (PCBs) in soils and plants. Environ. Pollut. Ser. A 28: 291-312.
6. MURPHY, T.J., J.C. POKOJOWCZYK, M.D. MULLIN. 1983. Vapor exchange of PCBs with Lake Michigan: the atmosphere as a sink for PCBs. In Physical Behavior of PCBs in the Great Lakes. (D. Mackay, S. Paterson, S.J. Eisenreich, M.S. Simmons, eds.), Ann Arbor Science, Ann Arbor, Michigan, pp. 49-58.
7. EISENREICH, S.J., B.B. LOONEY. 1983. Evidence for the atmospheric flux of polychlorinated biphenyls to Lake Superior, ibid., pp. 141-156.
8. MACKAY, D., P.J. LEINONEN. 1975. Rate of evaporation of low-solubility contaminants from water bodies to atmosphere. Environ. Sci. Technol. 9: 1178-1180.
9. BIDLEMAN, T.F., C.E. OLNEY. 1974. Chlorinated hydrocarbons in the Sargasso Sea atmosphere and surface water. Science 183: 516-518.
10. HARVEY, G.R., W.G. STEINHAUER. 1974. Atmospheric transport of polychlorobiphenyls to the North Atlantic. Atmos. Environ. 8: 777-782.

11. GIAM, C.S., E. ATLAS, H.S. CHAN, G.S. NEFF. 1980.
    Phthalate esters, PCB and DDT residues in the Gulf
    of Mexico atmosphere. Atmos. Environ. 14: 65-69.

12. ATLAS, E., C.S. GIAM. 1981. Global transport of
    organic pollutants: ambient concentrations in the
    remote marine atmosphere. Science 211: 163-165.

13. MACLEOD, K.E. 1981. Polychlorinated biphenyls in
    indoor air. Environ. Sci. Technol. 15: 926-928.

14. GSCHWEND, P.M., S. WU. 1985. On the constancy of
    sediment-water partition coefficients of hydro-
    phobic organic pollutants. Environ. Sci. Technol.
    19: 90-96.

15. LICHTENSTEIN, E.P. 1959. Absorption of some
    chlorinated hydrocarbon insecticides from soils
    into various crops. J. Agric. Food Chem. 7:
    430-433.

16. BEALL, M.L. JR., R.G. NASH. 1971. Organochlorine
    insecticide residues in soybean plant tops: root
    vs. vapor sorption. Agron. J. 63: 460-464.

17. BEALL, M.L. JR., R.G. NASH. 1972. Insecticide
    depth in soil: effect on soybean uptake in the
    greenhouse. J. Environ. Qual. 1: 283-288.

18. KLEIN, W., I. WEISGERBER. 1976. PCBs and environ-
    mental contamination. Environ. Qual Saf. 5:
    237-250.

19. THOMAS, W. 1979. Monitoring organic and inorganic
    trace substances by epiphytic mosses - a regional
    pattern of air pollution. In Proceedings of the
    13th International Conference on Trace Substances
    in Environmental Health, Columbia, pp. 285-289.

20. THOMAS, W., R. HERRMANN. 1980. Detection of
    chloropesticides, PCB, PCA (polycyclic aromatic
    hydrocarbons) and heavy metals in epiphytic mosses
    which acted as biofilters along a profile in
    central Europe. Staub-Reinhalt. Luft 40: 440-444.

21. THOMAS, W., H. SIMON, Å. RÜHLING. 1985. Classifica-
    tion of plant species by their organic (PAH, PCB,
    BHC) and inorganic (heavy metals) trace pollutant
    concentrations. Sci. Total Environ. 46: 83-94.

22. WSZOLEK, P.C., C.L. SCHOFIELD, D.J. LISK. 1980.
    Polychlorinated biphenyls in mosses in the
    Adirondack region of New York. N.Y. Fish Game J.
    27: 96-98.

23. GARTY, J., A.S. PERRY, J. MOZEL. 1982. Accumulation
    of polychlorinated biphenyls (PCBs) in the
    transplanted lichen Ramalina duriaei in air

quality biomonitoring experiments. Nord. J. Bot.
2: 583-586.

24. VILLENEUVE, J.-P., E. HOLM. 1984. Atmospheric
background of chlorinated hydrocarbons studied
in Swedish lichens. Chemosphere 13: 1133-1138.

25. BADSHA, K., G. EDULJEE. 1986. PCB in the U.K.
environment - a preliminary survey. Chemosphere
15: 211-215.

26. BUCKLEY, E.H. 1983. Decline of background PCB
concentrations in vegetation in New York State.
Northeastern Environ. Sci. 2: 181-187.

27. SLADE, D.H. 1968. Meteorological fundamentals for
atmospheric transport and diffusion studies. In
Meteorology and Atomic Energy 1968. (D.H. Slade,
ed.), U.S. Atomic Energy Commission, Office of
Information Services, NTIS, Springfield, Virginia,
pp. 13-63.

28. BUCKLEY, E.H. 1982. Accumulation of airborne
polychlorinated biphenyls in foliage. Science
216: 520-522.

29. FEDERAL REGISTER. 1979. Polychlorinated biphenyls
(PCBs): reduction of tolerances. Federal
Register 44: 38339.

30. FRIES, G.F. 1972. PCB residues: their significance
to animal agriculture. Agric. Sci. Rev. 10:
19-24.

31. FRIES, G.F., G.S. MARROW JR., C.H. GORDON. 1973.
Long-term studies of residue retention and
excretion by cows fed a polychlorinated biphenyl
(Aroclor 1254). J. Agric. Food Chem. 21: 117-121.

32. FRIES, G.F. 1982. Potential polychlorinated
biphenyl residues in animal products from application of contaminated sewage sludge to land. J.
Environ. Qual. 11: 14-20.

33. MOZA, P., I. WEISGERBER, W. KLEIN, F. KORTE. 1973.
Distribution and metabolism of carbon-14-labelled
2,2'-dichlorobiphenyl in the higher marsh plant
Veronica beccabunga. Chemosphere 5: 217-222.

34. MOZA, P., I. WEISGERBER, W. KLEIN, F. KORTE. 1974.
Metabolism of 2,2'-dichlorobiphenyl-^{14}C in two
plant-water-soil systems. Bull. Environ. Contam.
Toxicol. 12: 541-546.

35. MOZA, P., L. KILZER, I. WEISGERBER, W. KLEIN. 1976.
Contributions to ecological chemistry. CXV.
Metabolism of 2,5,4'-trichlorobiphenyl-^{14}C and
2,4,6,2',4'-pentachlorobiphenyl-^{14}C in the marsh

plant <u>Veronica beccabunga</u>. Bull. Environ. Contam.
Toxicol. 16: 454-463.

36. MOZA, P., I. WEISGERBER, W. KLEIN. 1976. Fate of
2,2'-dichlorobiphenyl-^{14}C in carrots, sugar beets,
and soil under outdoor conditions. J. Agric.
Food Chem. 24: 881-885.

37. MOZA, P., I. SCHEUNERT, W. KLEIN, F. KORTE. 1979.
Studies with 2,4',5-trichlorobiphenyl-^{14}C in
carrots, sugar beets and soil. J. Agric. Food
Chem. 27: 1120-1124.

38. MOZA, P., I. SCHEUNERT, W. KLEIN, F. KORTE. 1979.
Long-term uptake of lower chlorinated biphenyls
and their conversion products by spruce trees
(<u>Picea abies</u>) from soil treated with sewage
sludge. Chemosphere 6: 373-375.

39. MAASS, W.S.G., O. HUTZINGER, S. SAFE. 1975.
Metabolism of 4-chlorobiphenyl by lichens. Arch.
Environ. Contam. Toxicol. 3: 470-478.

40. PAL, D., J.B. WEBER, M.R. OVERCASH. 1980. Fate of
polychlorinated biphenyls (PCBs) in soil-plant
systems. Residue Rev. 74: 45-98.

41. WEBER, J.B., E. MROZEK JR. 1979. Polychlorinated
biphenyls: phytotoxicity, absorption and translocation by plants, and inactivation by activated
carbon. Bull. Environ. Contam. Toxicol. 23:
412-417.

42. AMICO, V., G. ORIENTE, M. PIATTELLI, C. TRINGALI.
1979. Concentrations of PCBs, BHCs and DDTs
residues in seaweeds of the east coast of Sicily.
Mar. Pollut. Bull. 10: 177-179.

43. MOUVET, C., M. GALOUX, A. BERNES. 1985. Monitoring
of polychlorinated biphenyls (PCBs) and hexachlo-
rocyclohexanes (HCH) in freshwater using the
aquatic moss <u>Cinclidotus danubicus</u>. Sci. Total
Environ. 44: 253-267.

44. YAMAMOTO, K., H. YAMAMOTO, T. MIZUTANI. 1982.
Uptake of 4-chlorobiphenyl and 4,4'-dichlorobi-
phenyl in six species of plant tissue cultures.
Bull. Environ. Contam. Toxicol. 28: 728-732.

45. MOSSER, J.L., N.S. FISHER, T.-C. TENG, C.F. WURSTER.
1972. Polychlorinated biphenyls: toxicity to
certain phytoplankters. Science 175: 191-192.

46. MOSSER, J.L., N.S. FISHER, C.F. WURSTER. 1972.
Polychlorinated biphenyls and DDT alter species
composition in mixed cultures of algae. Science
176: 533-535.

47. FISHER, N.S., C.F. WURSTER. 1973. Individual and combined effects of temperature and polychlorinated biphenyls on the growth of three species of phytoplankton. Environ. Pollut. 5: 205-212.

48. MOSSER, J.L., T.-C. TENG, W.G. WALTHER, C.F. WURSTER. 1974. Interactions of PCBs, DDT and DDE in a marine diatom. Bull. Environ. Contam. Toxicol. 12: 665-668.

49. FISHER, N.S. 1975. Chlorinated hydrocarbon pollutants and photosynthesis of marine phytoplankton: a reassessment. Science 189: 463-464.

50. O'CONNORS, H.B. JR., C.F. WURSTER, C.D. POWERS, D.C. BIGGS, R.G. ROWLAND. 1978. Polychlorinated biphenyls may alter marine trophic pathways by reducing phytoplankton size and production. Science 201: 737-739.

51. BIGGS, D.C., R.G. ROWLAND, C.F. WURSTER. 1979. Effects of trichloroethylene, hexachlorobenzene and polychlorinated biphenyls on the growth and cell size of marine phytoplankton. Bull. Environ. Contam. Toxicol. 21: 196-201.

52. NAU-RITTER, G.M., C.F. WURSTER, R.G. ROWLAND. 1982. Polychlorinated biphenyls (PCBs) desorbed from clay particles inhibit photosynthesis by natural phytoplankton communities. Environ. Pollut. Ser. A 28: 177-182.

53. POWERS, C.D., G.M. NAU-RITTER, R.G. ROWLAND, C.F. WURSTER. 1982. Field and laboratory studies of the toxicity to phytoplankton of polychlorinated biphenyls (PCBs) desorbed from fine clays and natural suspended particulates. J. Great Lakes Res. 8: 350-357.

54. GLOOSCHENKO, V., W. GLOOSCHENKO. 1975. Effect of polychlorinated biphenyl compounds on growth of Great Lakes phytoplankton. Can. J. Bot. 53: 653-659.

55. MAHANTY, H.K., B.A. FINERAN, P.M. GRESSHOFF. 1983. Effects of a polychlorinated biphenyl (Aroclor 1242) on the ultrastructure of certain planktonic algae. Bot. Gaz. 144: 56-61.

56. UREY, J.C., J.C. KRICHER, J.M. BOYLAN. 1976. Bioconcentration of four pure PCB isomers by Chlorella pyrenoidosa. Bull. Environ. Contam. Toxicol. 16: 81-85.

57. MAHANTY, H.K., B.A. FINERAN. 1976. Effects of a polychlorinated biphenyl (Aroclor 1242) on the

ultrastructure of frond cells in the aquatic
plant Spirodela oligorrhiza (Kurz) Hegelm.
N.Z. J. Bot. 14: 13-18.

58. SINCLAIR, J., S. GARLAND, T. ARNASON, P. HOPE, M.
    GRANVILLE. 1977. Polychlorinated biphenyls
    and their effects on photosynthesis and respiration.
    Can. J. Bot. 55: 2679-2684.

59. STREK, H.J., J.B. WEBER, P.J. SHEA, E. MROZEK JR.,
    M.R. OVERCASH. 1981. Reduction of polychlorinated
    biphenyl toxicity and uptake of carbon-14 activity
    by plants through the use of activated carbon. J.
    Agric. Food Chem. 29: 288-293.

60. STREK, H.J., J.B. WEBER. 1982. Adsorption and
    reduction in bioactivity of polychlorinated
    biphenyl (Aroclor 1254) to redroot pigweed by
    soil organic matter and montmorillonite clay.
    Soil Sci. Soc. Am. J. 46: 318-322.

61. KLEKOWSKI, E.J. JR., E. KLEKOWSKI. 1982. Mutation
    in ferns growing in an environment contaminated
    with polychlorinated biphenyls. Amer. J. Bot.
    69: 721-727.

62. MROZEK, E. JR., W.H. QUEEN, L.L. HOBBS. 1983.
    Effects of polychlorinated biphenyls on growth
    of Spartina alterniflora Loisel. Environ. Exp.
    Bot. 23: 285-292.

62a. REINHOLTZ, W., A.A. VOLPE. 1977. Elimination of
     polychlorinated biphenyl pollutants from water by
     means of aquatic plants. In Abstract of Papers,
     173rd ACS Meeting, New Orleans, Louisiana, March
     20-25, 1977, ACSC Abstract #41.

63. IWATA, Y., F.A. GUNTHER, W.E. WESTLAKE. 1974.
    Uptake of a PCB (Aroclor 1254) from soil by
    carrots under field conditions. Bull. Environ.
    Contam. Toxicol. 11: 523-528.

64. WALLNÖFER, P., M. KÖNIGER, G. ENGELHARDT. 1975.
    Fate of xenobiotic chlorinated hydrocarbons (HCB
    and PCBs) in plants and soils. Z. Pflanzenkr.
    Pflanzenschutz 82: 91-100.

65. SAWHNEY, B.L., L. HANKIN. 1984. Plant contamination
    by PCBs from amended soils. J. Food Prot. 47:
    232-236.

66. SUZUKI, M., N. AIZAWA, G. OKANO, T. TAKAHASHI. 1977.
    Translocation of polychlorobiphenyls in soil into
    plants: a study by a method of culture of soybean
    sprouts. Arch. Environ. Contam. Toxicol. 5:
    343-352.

67. MROZEK, E. JR., E.D. SENECA, L.L. HOBBS. 1982.
    Polychlorinated biphenyl uptake and translation by
    Spartina alterniflora Loisel. Water, Air, Soil
    Pollut. 17: 3-15.
68. MROZEK, E. JR., R.B. LEIDY. 1981. Investigation of
    selective uptake of polychlorinated biphenyls by
    Spartina alterniflora Loisel. Bull. Environ.
    Contam. Toxicol. 27: 481-488.
69. MARINUCCI, A.C., R. BARTHA. 1982. Biomagnification
    of Aroclor 1242 in decomposing Spartina litter.
    Appl. Environ. Microbiol. 44: 669-677.
70. BUSH, B., L.A. SHANE, L.R. WILSON, E.L. BARNARD. D.
    BARNES. 1986. Uptake of polychlorobiphenyl
    congeners by purple loosestrife (Lythrum salicaria)
    on the banks of the Hudson River. Arch. Environ.
    Contam. Toxicol. 15: 285-290.
71. WALLNÖFER, P., M. KÖNIGER. 1974. Model experiments
    on the uptake of hexachlorobenzene and polychlori-
    nated biphenyls by cultivated plants from different
    substrates. Nachrichtenbl. Dtsch. Pflanzenschutz-
    dienst (Braunschweig) 26: 54-57.
72. IWATA, Y., F.A. GUNTHER. 1976. Translocation of the
    polychlorinated biphenyl Aroclor 1254 from soil
    into carrots under field conditions. Arch.
    Environ. Contam. Toxicol. 4: 44-59.
73. JACOBS, L.W., S.-F. CHOU, J.M. TIEDJE. 1976. Fate of
    polybrominated biphenyls (PBBs) in soils. Persis-
    tence and plant uptake. J. Agric. Food Chem. 24:
    1198-1201.
74. FRIES, G.F., G.S. MARROW. 1981. Chlorobiphenyl
    movement from soil to soybean plants. J. Agric.
    Food Chem. 29: 757-759.
75. BACCI, E., C. GAGGI. 1985. Polychlorinated biphenyls
    in plant foliage: translocation or volatilization
    from contaminated soils? Bull. Environ. Contam.
    Toxicol. 35: 673-681.
76. BIDLEMAN, T.F., C.E. OLNEY. 1974. High-volume collec-
    tion of atmospheric polychlorinated biphenyls.
    Bull. Environ. Contam. Toxicol. 11: 442-450.

Chapter Eight

CHEMICAL INTERACTIONS OF ACIDIC PRECIPITATION AND
TERRESTRIAL VEGETATION

LANCE S. EVANS

Laboratory of Plant Morphogenesis
Manhattan College
The Bronx, New York 10471

203

INTRODUCTION

Acidic precipitation — wet or frozen precipitation
with a H^+ concentration greater than 2.5 ueq liter^{-1}
(equivalent to a pH of about 5.6) — is a significant air
pollution problem in North America and Europe. The
northeastern portion of the United States is at the center
of the high acidic rainfall area in North America.[1] The
high H^+ concentration of precipitation (rain, snow, fog,
sleet, and mist) in the northeastern United States is
explained by the presence of strong acids. Sulfuric acid
contributes a portion of the acidity,[2-4] and nitrate and
chloride are significant anion components of the total
acidity in precipitation.[5,6] A significant amount of
sulfur dioxide emitted into the atmosphere is converted
into sulfuric acid and various aerosols of ammonium
sulfate. Particulate sulfur compounds and sulfur oxides
may be incorporated into precipitation with conversion to
H_2SO_4.[2] About 90% of the sulfur in the atmosphere of the
northeastern United States is contributed by antrhopogenic
sources.[7] An estimate for nitrogen inputs into the
atmosphere of the Adirondack Mountain Region indicates
that 34% of the anions in rain could be attributed to
nitrates.[8] Acidic precipitation is only a portion of the
total acidity brought to the earth's surface from the
atmosphere. Dry fall plus acidic precipitation is termed
acidic deposition.

This paper will review pertinent information on the
chemical interaction of acidic precipitation and plant
foliage. It will demonstrate that acidic deposition
decreases yields of field-grown soybeans, that acidic
deposition decreases the protein contents of harvested
seeds of some cultivars of field-grown soybeans and that
these results occur without any visible foliar injury
(no detectable changes in anatomy and morphology) which
suggests that acidic precipitation causes these effects
by altering the chemical environment. The second section
describes the chemical and physical characteristics of
precipitation and plant foliar surfaces, the first site
of interaction between precipitation and vegetation. The
third section describes the interaction of acidic
precipitation and plant surfaces in terms of changes in
plant surface characteristics by acidic deposition,
changes in precipitation chemistry by leaf surfaces,
incorporation of materials from rain into vegetation,

changes in leaf cell permeability of leaves exposed to
rainfall acidity, and removal of materials from foliage by
rain. Finally, this chapter will speculate on how these
chemical interactions of acidic deposition and plant
surfaces may account for, at least in part, the decreases
in soybean seed yields and seed quality and other
possible vegetative effects. Hypotheses will be
proposed to test this speculation.

"INVISIBLE" INJURY TO TERRESTRIAL VEGETATION

Yield Effects To Soybeans

        Based upon research under controlled environmental
conditions, it appears that the foliage of broad-leaved
herbaceous plants is most sensitive to acidic rainfalls.[9]
Results of a preliminary field experiment performed in
the summer of 1979 suggested that field-grown soybeans,
a broad-leaved herbaceous crop, was sensitive to acidic
deposition.[10] These results prompted a series of
experiments with field-grown soybeans that commenced in
1981 and employed moveable rainfall exclusion shelters.

        Moveable rainfall exclusion shelters (BNL-RES) were
first conceived and constructed at the Brookhaven National
Laboratory, Upton, New York and used successfully there
from 1981 through 1985. The advantages of BNL-RES
systems are: (1) the moveable exclusion shelters move
over the experimental area only when ambient rainfalls
occur or simulated rainfalls are applied; (2) the plant's
microclimate is altered minimally; (3) the chemical,
physical, temporal, and spatial aspects of the application
of rainfalls used to simulate a variety of acidity
exposures can be regulated precisely. Since the mean
weekly amount of total water deposition was applied each
year, results among years can be compared.

        Results of the soybean field experiments at Upton
from 1981 through 1984 have been published;[11-14] results of
the 1985 experiments are not yet published but are similar
to those of 1984. The studies indicate that seed yields
of the Amsoy 71 cultivar of soybeans are reduced by
exposure to increasingly acidic simulated rainfalls.
These reductions are reflected in the number of seeds
per plant, the mass of seeds per plant, and the seed

yield per unit area. The decrease in soybean seed yield
was attributed largely to a decrease in the mean number of
mature pods per plant, since other factors which could
influence yield exhibited less pronounced trends which
did not parallel the significant changes observed for the
measures of yield mentioned above. These factors include
seed number per pod, mass per individual seed, and plant
population density. A decrease in number of pods per
plant may result from a reduction in flower production,
pollination, or fertilization, a decrease in pod retention,
and/or inadequate pod development.

Asgrow 3127, Corsoy, and Hobbit soybean cultivars
showed negative responses to simulated rainfall acidity
in a manner similar to Amsoy. The negative responses of
these three cultivars was not as large as that experienced
by the Amsoy cultivar. For example, in a 1984 field
experiment, Asgrow plants shielded from ambient rainfalls
by BNL-RES and exposed to simulated rainfalls of pH 4.4,
4.1, and 3.3 exhibited respective yields of 14.5, 12.2,
and 9.0% below the yields of plants exposed to simulated
rainfalls of pH 5.6. For Corsoy, comparable figures were
13.7, 12.7, and 7.8% and for Hobbit the figures were 9.2,
6.2, and 16.6%, respectively.[14]

While these results, combined with those of comparable
experiments at Urbana, Illinois show that several varieties
of soybeans are sensitive to rainfall acidity,[15] it should
be noted that the Williams cultivar of soybeans does not
appear to be sensitive to rainfall acidity.[12,15] Other
soybean varieties, grown under different experimental
conditions have also been reported to be insensitive to
rainfall acidity.[16,17]

Qualitative observations on the condition of the
plants in the experiments described above were also
recorded. Foliar injury was observed on young plants
exposed to simulated rainfalls of pH 2.7 or 3.3. Plants
exposed to simulated rainfalls of pH 4.1 or 5.6 exhibited
no visible foliar injury attributable to rainfall acidity.

Bean yields for the Amsoy cultivar shielded from
ambient rainfall and exposed to twice weekly simulated
rainfalls of pH 5.6 were consistent for the growing seasons
of 1981 (13.1 g per plant; 4660 kg ha^{-1}); 1982 (13.3 g
per plant; 4590 kg ha^{-1}); and 1983 (13.0 g per plant;

4260 kg ha^{-1}). These relatively consistent yearly yields
can be contrasted with the more variable yields of plants
exposed to ambient rainfalls. Yields of Amsoy plants
exposed to ambient rainfalls for 1981, 1982 and 1983 were
11.7 g per plant (4460 kg ha^{-1}), 11.4 g per plant (4000
kg ha^{-1}), and 17.6 g per plant (4820 kg ha^{-1}), respectively. Differences in the yields of plants grown in
successive years under ambient conditions may be attributed
in part to yearly weather variations. For instance,
during the three growing seasons mentioned above, ambient
rainfalls totalled 212, 457 and 361 mm, respectively.

The protocol used in these experiments (exposing the
plants to simulated rainfalls based on long term average
ambient occurrences) provides yeild responses under average
rainfall conditions. It is recognized that this benefit
is gained at the cost of collecting any information about
the range of responses which may occur under widely
varying exposures to rainfall within and among growing
seasons. Other experiments, with a different emphasis,
will need to be employed to examine these variables.

Moisture availability and utilization have considerable effects on plant growth and reproduction. The
results of a soybean field experiment in 1983 indicate
that the productivity of plants exposed to twice-weekly
simulated rainfalls of longer duration was greater than
with daily rainfalls of shorter duration even though the
plants exposed to daily rainfalls received more water.
These differences probably resulted from the fact that
the long duration rainfall regimen more effectively
supplied moisture to plant roots. It is noteworthy that
even though the yield differences between the two treatment frequencies were greater than the effect of rainfall
acidity, the slopes of the dose-response functions for
the two frequencies for each cultivar were markedly
similar. These results show that simulated rainfall
frequency and duration can significantly affect soybean
yields independently of rainfall acidity effects at the
frequencies and durations tested.

Altered Protein Content of Soybean Seeds

Protein analyses of seeds harvested from the various
soybean field experiments show that the protein content of
such seeds can be reduced when plants are exposed to high

levels of rainfall acidity. In the field experiments of
1979 and 1981, statistically significant decreases in
protein content were observed.[18] However, in the 1982
experiment no significant differences in seed protein
content resulted.[18]

In an experiment conducted in 1983, seed yields of
plants exposed twice weekly to long duration rainfalls
(70 min, twice weekly) showed no significant differences
in seed protein contents among the acidity levels tested.
In contrast, significant differences in protein contents
were found for seed samples from plants exposed to daily
rainfalls of short duration (21 min). Seeds of plants
exposed to simulated rain of pH 5.6 had average protein
contents of 37.9% compared with a range between 30.6 and
32.1% from plants exposed to lower pH values. Overall
the protein contents of seeds of acidic rainfall treatments, expressed on a per plant basis, ranged from 23
to 34%. The results suggest that the response of soybean
seed yields to acidity is affected by the duration and
frequency of simulated rainfalls.

In the 1984 growing season, four commercial cultivars
were tested under conditions of longer duration rainfalls
(70 min, twice weekly). Significant effects were
observed only with the Amsoy cultivar and not with
Asgrow, Corsoy or Hobbit. For the latter three cultivars
the protein contents of samples were all relatively high
(the means were between 35 and 39%). The relatively low
protein content of Amsoy soybeans exposed to simulated
rainfalls of pH 2.7 was apparently responsible for the
statistical significance among treatments.[19]

Several generalizations are evident from the data of
the seven experiments. First, the results show that the
yields of both seed and seed protein are generally lower
in plants exposed to high acidity than in those exposed
to low acidity. Second, decreases in protein content as
a result of increased rainfall acidity are greatest when
expressed on a per plant basis. Third, seed yields and
protein contents vary from year to year. Finally, the
effects of rainfall acidity on seed yields are independent
of seed protein content. This latter statement appears
true because, for example, in the rainfall experiment of
1981 the decrease in protein content per plant was due
mostly to changes in protein content per seed mass, while

for the rainfall exclusion experiment of 1982 the decrease was due to differences in seed mass per plant.

From available evidence it appears that the protein content seed of Amsoy soybeans are sensitive to simulated rainfall acidity. In contrast, results of one year of experimentation with Asgrow 3127, Corsoy 79, and Hobbit suggest that the protein content seeds of these cultivars are not sensitive to simulated rainfall acidity. These different responses to acidity among cultivars are similar to seed yield responses to rainfall acidity.[19] Although differences in protein content and seed yields among cultivars of soybeans occur, the underlying reasons for differences in responses remain to be addressed.

Lack of Visible Foliar Injury in Field Experiments

In the experiments described above in which soybean seed yields and protein contents were negatively affected by simulated acidic rainfalls, foliar injury was observed only on young plants shielded from ambient rainfalls and exposed to simulated rainfalls of pH 3.3 and pH 2.7. Visible foliar injury was never present on plants exposed to rainfalls of pH 4.1 and above. Moreover, no visible foliar injury occurred to any plants following the third week after germination.

The appearance of the small amount of visible injury to young plants exposed to high acidity rainfall did not result in growth stunting or any other visible symptom later in the growing season. This lack of any lasting visible injury together with the consistent changes in yield suggest that the decreases in soybean yields do not occur by cell destruction but by changes in cellular, tissue, or whole plant metabolism. Observations such as these have lead researchers to question if changes in mineral nutrients caused by rainfall acidity may result in changes in plant productivity.

CHEMICAL AND PHYSICAL NATURE OF ACIDIC PRECIPITATION AND PLANT SURFACES

Chemical and Physical Nature of Acidic Precipitation

The chemical and physical characteristics of precipitation are complex and diverse. Many factors interact to produce deposition to the earth's surface. At the same time, a large number of chemical species are involved.

Chemical Nature of Precipitation. The chemical composition of rainfall in various geographic regions has been determined.[20-22] Moreover, various estimates of amounts of SO_2 and sulfate deposited on terrestrial ecosystems have been made. One estimate is that 200 teragrams (1 teragram = Tg = 10^{12} g) of sulfur move from the earth's surface to the atmosphere and back again annually.[23] About 15 Tg are absorbed by vegetation directly as SO_2, while about 20 Tg (10%) and 80 Tg (40%) are deposited as sulfates on terrestrial ecosystems by dry and wet deposition, respectively.[23] A similar estimate of annual sulfur deposition was obtained independently.[24] Much of this sulfur impacts upon land areas of the Northern Hemisphere. About 93% of all anthropogenic sulfur production occurs in the Northern Hemisphere and about 80% of this total is deposited on land surfaces in that hemisphere.[24]

Although dry deposition of sulfur compounds is significant and these compounds may have some phytotoxic effects at high concentrations, about one-half of atmospheric sulfur reaches the earth's surface in precipitation (wet deposition). Indeed, most of this sulfur is in the form of sulfate with associated H^+ and NH_4^+ ions.[3] Wet deposition may consist of rain, snow, and fog. Although fog may constitute a significant portion of wet deposition in some areas, it usually does not result in large amounts of moisture input to foliage in most areas of the world. There are no estimates of the total amount of acidity deposited as dew or fog on vegetation.

Data from the National Atmospheric Deposition Program (NADP) for 1978 and 1979 show that the median pH of precipitation in portions of New York, Ohio, and Pennsylvania is less than 4.2, while most of the northeastern United States has a median pH below 4.4.[25] pH levels derived from NADP are in good agreement with data of the MAP3S and EPRI networks.[26]

Acidity in precipitation events can be categorized by the percentage of rainfalls within various pH ranges. The frequency distribution of the pH of ambient rainfalls during the 1981 growing season on an event basis at Upton, New York provides a good example.[11] Values are weighed volume means of hourly samples collected with a sequential rain sampler for the period 30 May through 29 September 1981. A majority (55%) of samples had pH levels between 3.5 and 4.5 and about 80% of all samples were between pH 3.5 and 5.0. No rainfall samples had a mean pH below 3.0 or above 7.0. The volume weighed mean H^+ concentration of all rainfalls was 91.5 ueq liter^{-1} (pH 4.04). Data from rain samples collected for the years 1976-1979 at MAP3S stations at Ithaca, New York; University Park, Pennsylvania; and Charlottesville, Virginia were chosen to be compared with data from Upton, New York because these are three representative locations within the northeastern United States.[1] Most rainfall had pH levels between pH 3.5 and 4.5. Thirty-two percent of all rainfall samples had pH values below 4.0. Less than 0.5% of all rainfall was below pH 3.0 or above pH 5.5. These data demonstrate that a wide diversity in rainfall chemistry is present at several locations and that presenting only long-term (i.e., annual) volume weighed mean pH values may obscure this diversity.

Most low pH values in rainwater occur in the spring and summer.[27] The mean (weighed) hydrogen ion concentration in rainfall at Brookhaven National Laboratory was almost 10^{-4} M (pH 4.0) during the summer of 1977. Values for spring were about 0.45 X 10^{-4} M (pH 4.35) and the mean concentrations for winter and fall were similar at about 0.23 X 10^{-4} M (pH 4.64). These observations indicate that acidity of rainfall is highest when most plants are actively growing.

The chemical composition of ambient rainfall varies markedly not just with respect to acidity but with regard to total ion composition.[1] For example, results of long-term sampling in the New York area show that mean concentrations (ueq ℓ^{-1}) of ionic constituents present in rainfall were: ammonium, 27.96; sodium, 45.63; calcium, 8.3; magnesium, 11.4; potassium, 10.0; iron, 3.22; zinc, 3.98; nickel, 1.70; copper, 5.98; cadmium, 0.71; lead, 12.35; hydrogen ion, 71.7; manganese, 1.09; sulfate, 96.93; nitrate, 49.06; chloride, 51.80; and

fluoride, 5.26.[28] The values for sodium and chloride were
obtained from samples taken at Upton, New York only and
may be higher than a regional mean because of the coastal
influence.

Frequency and Duration of Ambient Rainfall. Data
from national or international precipitation chemistry
programs are derived from precipitation collectors that
collect wet precipitation over periods of days, weeks,
and months. Recent results suggest that most ambient
rainfall during the summer of 1982 at field site at
Brookhaven National Laboratory were of short duration and
were of low precipitation rate. Thirty-six precipitation
events (33% of all events) had durations of less than 2
minutes and 47 events (43%) were shorter than 10 minutes.
Forty-two events (38%) were between 11 and 140 minutes
while the remaining 19 were longer than 140 minutes. A
majority of events had rates less than 1.5 mm hr^{-1} and
only 12.5% of all events had rates greater than 3.5 mm
hr^{-1}. Over 50% of all ambient showers during the summer
period had volumes below 0.25 mm. Moreover, only 8% of
all events had amounts greater than 8 mm.[12] These data,
derived from a rotary rain indicator and a tipping bucket
guage, indicate that the majority of all ambient rainfall
events are of relatively short duration and small
volume.[10]

Data were analyzed from other locations in the
United States to determine if the frequency and duration
of ambient rainfall were similar to those described
above from Long Island. The data were taken during the
crop growing season (1 May through 30 September)[29] on a
minute-by-minute basis with 0.0042 mm per minute volume
resolution. A significant percentage (between 17 and
40%) of all showers recorded at Urbana, Illinois; Franklin,
North Carolina; and Seaside Park, New Jersey, had durations
of less than 20 minutes. Similarly, a large percentage
had durations of less than 40 minutes (50, 46, and 32% for
3 years at Urbana, Illinois; 65% at Coral Gables, Florida;
50% at Franklin, North Carolina; and 54% at Seaside Park,
New Jersey). When all data from the six growing seasons
were pooled, over 88% of all rainfalls were shorter than
160 minutes.

Similar results were observed when volume measurements
were made. Between 38 and 57% of all recorded showers had

less than 1 mm at the five stations in the eastern United
States listed above. When all data were pooled, over
65% of all rainfalls were less than 3 mm in volume.
These data demonstrate that plant foliage is wetted by
numerous showers which are of short duration and low
volume. These data also show that taking samples over
periods of days, weeks, and months to determine the
chemical characteristics of rainfall may obscure a large
diversity of characteristics that plant foliage may
experience.

Chemical and Physical Nature of Plant Surfaces: Relation
to Surface Solutions

     The surfaces of terrestrial vegetation are covered
with a cuticle and epicuticular waxes of different
chemical and physical characteristics. The significance
of these differences in characteristics is probably as
varied as the climatic and microclimatic environments
under which plants must survive.

     Foliage of leaves and herbaceous stems are covered
by the plant cuticle. The cuticle is composed of epicuticular waxes on the surface overlaying a cuticularized
layer which is composed of cellulose and pectin encrusted
with cutin. Inside this cuticularized layer a cutinized
layer of the cell wall is present. Lastly, a pectin
layer is located between the cutinized cell wall and the
plasmalemma of epidermal cells.[30]

     A large variety of epicuticular waxes is present on
surfaces of foliage. Many n-primary alcohols and n-fatty
acids varying in chain length from 24 to 34 carbon atoms
have been characterized. Other waxes such as n-nonacosane,
n-hextriacontane, n-hexacosan-1-ol, and other substituted
alcohols have been detected. Nonacosane $(CH_3-(CH_2)_{13}-CH_2-$
$(CH_2)_{13}-CH_3)$ results from a condensation of 2 molecules of
pentadecanoic acid, and subsequent decarboxylation to form
nonacosan-15-one followed by reduction. Many other final
product waxes are possible when different fatty acids are
subjected to condensation.

     Cutin is a large polymer of fatty acids. Leaves
contain three fatty acid oxidizing enzymes for cutin
synthesis from fatty acids: lipoxidase, stearic acid
oxidase, and oleic acid oxidase. There is evidence that

stearic acid, produced from acetate, is converted to
oleic acid by stearic acid oxidase. In turn, oleic acid
is converted to linoleic acid by oleic acid oxidase.
Linoleic acid (and possibly linolenic acid) are oxidized
to their hydroperoxides by lipoxidase. In turn, these
hydroperoxides can react with a variety of fatty acids to
form saturated and unsaturated hydroxy-fatty acid
condensation products.[30]

To produce the entire cutin polymer, peroxide bonds
are formed between fatty acids and hydroxy-fatty acids.
After these reactions, esterifications between hydroxyl
and carboxyl groups can easily occur. The final polymerization step probably involves other enzymes including
catalase that can form various alkoxy and hydroxy radicals
by splitting peroxide linkages. This polymerization
process is completed when repeated addition of other
chains occurs in a highly cross-linked pattern.[30]

Waxes and wax derivatives in plants vary greatly.
Many surface waxes contain n-alkanes (C_{13}-C_{34}), n-primary
alcohols (C_{16}-C_{32}), mono- and dimethylalkanes, some
alkenes, some fatty acids and other compounds.[30] Some
unique compounds include tricontane in Gossypium hirsutum,
n-tritriacontan-16-18-dione in several Eucalyptus species,
sterols from sugarcane (Saccharum officinarum), wax and
estolides of carnauba wax. Plant waxes frequently are
specific for the species and genus in which they occur.
Consequently they have been used extensively in plant
taxonomic studies which reinforces the wide diversity of
chemical structures of surface waxes.[30]

From available evidence it appears that cuticular
substances are secreted through the walls of epidermal
cells. In some cases, walls on all sides of cells become
cutinized to form a cuticular epithelium.[30] If large
cutin deposits occur in the middle lamellae, cutin
cystoliths may result.

Cuticles usually develop during early organ ontogeny.
The early formed cuticle hardens by oxidation and
continued polymerization through the addition of fatty
acids, alkanes, etc. In cuticles of young organs,
surface waxes are extruded through and within the
cuticular layer. From available evidence it appears as
if the cuticle develops through leaf expansion only and,

at least for most species, little or no cuticular development occurs after leaf expansion is completed. Young leaves of <u>Pisum sativum</u> and <u>Eucalyptus</u> species show prominent crystalline and tubular wax formation that last throughout ontogeny.[31]

The amount of injury to plant foliage by acidic precipitation may depend upon the area of leaves in contact with rainwater. Moreover, injury may also depend upon the rate of absorption of materials from rainwater per unit area. The amount of water absorbed by foliage depends upon many characteristics. These characteristics differ among plant species and, as a result, may determine relative species-sensitivity to precipitation acidity.

The amount of foliar injury may be a function of foliar wettability. The attraction of a liquid to a solid surface upon impaction is an important measure of the amount of water retained on the surface. However, the area over which a droplet spreads depends upon the relative amount of attraction between the liquid and the surface. The advancing contact angle has been used as a criterion of leaf surface wettability[30] and is defined as the angle between the surface of the leaf and the tangent plane of a water droplet at the circle of contact between air, liquid, and leaf. A zero degree angle would maximize wettability while an angle of 180° would provide virtually no wetting. Contact angles of pure water drops may be as small as 31° on <u>Phragmites communis</u> to greater than 150° on <u>Triticum aestivum</u> and <u>Lupinus albus</u>.[31-33] In this way, the contact angle is determined mostly by chemical and physical characteristics of epicuticular waxes and to a lesser extent by surface roughness.[30] Wettability of leaf surfaces increases markedly when the cuticle and epicuticular waxes are removed.[34]

Solutions must penetrate through cuticular layers or through stomata to reach leaf cells. It has been postulated that cuticles are perforated with micropores.[35] Cuticular pores may be numerous in specialized areas such as the bases of trichomes, hydathodes, glandular hairs,[36] water-absorbing scales of Bromeliaceae,[37] and stigmas.[38] Although cuticular pores may not be present, solutions penetrate faster at these locations.[30] These results may explain why injury from acidic precipitation occurs more frequently at bases of trichomes and hydathodes (see

below). Penetration of rain or leaf surface solutions
through stomata is thought to be infrequent if it occurs
at all.[39-42] Generally, spontaneous infiltration of
stomata by water will occur if the contact angle is
smaller than the angle of the aperture wall. The degree
of stomatal opening from 4 to 10 μm is of little importance
in penetration. However, cuticular ledges present at the
entrance to the outer vestibule and between the inner
vestibule and substomatal chamber resulted in very small
wall angles. These small wall angels may not be adequate
for water penetration of stomata. All evidence suggests
that the main route of entry of substances should be
through the cuticle and not through stomata.

Non-polar molecules should penetrate the cuticle at
rates faster than those of polar molecules because of the
non-polar nature of both cutin and epicuticular waxes.[44]
More polar compounds, such as inorganic ions and water,
may preferentially enter the leaf via pectinaceous
channels that traverse the cuticle.[45] It is postulated
that the ratio of cutin to wax in the cuticle influences
the degree of penetration of polar substances.[30]

Solution acidity may also affect cuticular penetration
rates.[46,47] In experiments with isolated cuticles of
apricot leaves, penetration rates of acidic substances
increased with an increase in solution acidity. The
penetration rates of basic substances increased with a
decrease in solution acidity. These relationships between
pH and sorption rates with apricot cuticles[48] have been
verified with isolated cuticles[49] and intact bean leaves
exposed to buffered solutions.[28]

Cuticles and epicuticular waxes should not be
considered to be inert, non-differential barriers to ion
permeability. Relative permeability coefficients of
isolated cuticles to various mono- and divalent cations
differ by as much as a factor of six. The idea of
differential or selective uptake of chemical constituents
by foliage is reinforced by other experimental data.
Penetration of chemicals into foliage is affected by both
metabolic and non-metabolic processes. Reversible
diffusion through cuticles is followed by metabolically-
controlled uptake through cellular membranes.[50] There are
many examples of light-enhanced foliar uptake of
elements[40,42,51] that is inhibited by protein synthesis

inhibitors.[52] In this way, uptake of various elements by foliage is selective to an unknown degree and is dependent upon the metabolism of leaf cells.

Some regions of the cuticle are more permeable than others. For example, basal portions of trichomes and guard cells[42,53] are preferential sites of absorption. Moreover, absorption of water-soluble materials may be rapid through the cuticle near veins.[33,54] These studies are germane because about 95% of all foliar lesions following exposure to simulated acidic rain under controlled conditions occurred at the bases of trichomes, at guard and subsidiary cells of stomata, and along veins.[9,55,59] From these data it seems reasonable to conclude that phytotoxic components of simulated acidic precipitation penetrate the cuticle at faster rates near vascular tissues, subsidiary cells, and at the bases of trichomes and hydathodes.

CHEMICAL INTERACTIONS OF ACIDIC PRECIPITATION AND TERRESTRIAL VEGETATION

Changes in Plant Surface Characteristics by Acidic Deposition

Plant cuticles and tissues are subject to physical and chemical weathering. Physical damage may occur by plant tissues rubbing on objects; by other leaves; by rain, hail, water splash; and by deposition of soot, oils, sand, and dust.[30] For example, numerous studies have shown that retention of herbicides increased after physical and chemical weathering. Recovery from excessive weathering may occur in expanding leaves in which cutin and wax deposits are still being occluded into the cuticle. It must be emphasized that cuticular weathering affects all surface properties of leaves. These results support the idea that plant growth and nutrient balance is the result of their entire environment.

Acidic precipitation may change the surface characteristics of foliage. It has been suggested that acidic precipitation may affect the submicroscopic structure of the epicuticular wax layers of leaves. Shriner[60] presented scanning electron micrographs that showed that leaves of kidney bean and willow oak exposed to simulated

rain of pH 3.2 had eroded superficial waxes, cutin, and
cuticular waxes. There was only slight erosion on leaves
exposed to rain of pH 6.0. In contrast, Paparozzi,[58] who
used scanning and transmission electron microscopy,
observed no erosion of epicuticular or cuticular waxes on
either yellow birch or kidney bean after exposure to
simulated rainfalls of pH as low as 2.8. The cuticular
waxes were not structurally changed by simulated acidic
rain even though underlying cells were affected. Moreover,
the most widely used methods of isolating cuticles involve
exposure to strong acids.[49,61] At the present a relationship between changes in the cuticle, including cutin and
epicuticular waxes, and phytotoxic changes in leaf cells
by acidic precipitation remains to be established.

Changes in Precipitation Chemistry by Leaf Surfaces

Leaf surfaces may also change the chemistry of rain
solutions. The acidity of rainwater may be changed by
chemicals from leaves as raindrops dry. Adams[59] demonstrated that leaves of several plant species have
differing buffering capacities. The pH of simulated
raindrops (50 µl) of pH 5.6, 3.5, and 3.0 usually
increased during an observation period of the 75 minutes.
However, when the initial pH was 2.5 the pH always
decreased with time. The pH of a parafilm surface (control)
was always lower than that of any of the plant surfaces
suggesting that leaves produced substances to neutralize
the acidity. The presence on the surfaces of substances
produced by leaves has been recognized for almost a
century.[62] Other studies by Evans et al.[63] substantiate
the results of Adams.[59] In these experiments pH measurements were obtained from 50 µl droplets of simulated rain
solutions. The volume of droplets decreased by about 30%
every 30 min during exposure periods of 130 to 160 min.
Droplet volumes were the same on leaflet surfaces as on
teflon plates and had a pH of 3.1 and 2.7, respectively.
Acidities of drops of pH 3.1 and 2.7 increased to pH 2.0
and 1.4, respectively, as volume reduction occurred.
During this period the number of picoequivalents in
dropments of pH 3.1 and 2.7 on leaflets decreased by 62
and 74%, respectively, over 120 min. During the same time
period the number of picoequivalents in droplets of pH 3.2
and 2.9 on teflon decreased by only 47 and 53%, respectively. The pH of droplets of initial pH 4.1 on leaflets
did not change significantly during droplet volume

reductions while the acidity of droplets on teflon
increased slightly. At pH 5.6, solution pH increased
slightly as volume decreased. The change in measured pH
and number of picoequivalents per droplet were markedly
similar on both leaves and teflon plates. Although they
provided similar data, these experiments represent only a
single exposure of a leaf surface to acidic conditions.
At present, no data on the effects of multiple exposures
is available that relate frequency of precipitation on the
same leaves to an acidity level. Possibly, very different
responses would be obtained if multiple exposures occurred.

Incorporation of Materials from Rain into Vegetation

Much research has demonstrated that the main route of
penetration of foreign materials in aqueous solutions is
through the cuticle.[30] In general, solution penetration
occurs by diffusion which is dependent upon three factors:
the time period of exposure, the surface area of contact,
and the chemical nature of the substance absorbed. The
first two of these factors are related to the degree of
leaf wettability described previously.

Penetration of substances into foliage occurs by a
combination of metabolic and non-metabolic processes.
Specifically, substances pass through the cuticle itself
by passive, reversible diffusion, but cellular uptake is
under metabolic control at the plasmalemma.[50] Evidence
demonstrates that the rate limiting step in uptake of
substances by foliage is metabolically controlled whether
by energy needs, protein carriers, or other metabolically
regulated processes.[52]

Since a large variety of chemicals in a wide range of
concentrations are found in ambient precipitation in the
eastern United States,[64-67] it is necessary to know if
materials in rain can be absorbed by foliage and if rainfall can remove materials from foliage. Recent information
suggests that materials can be absorbed at different rates
from rainfall of varying acidity.[28] For example, sulfate
anion penetrated leaves faster at pH 2.7 than at 5.7
while $^{86}Rb^+$ penetrated faster at pH 5.7 than at lower pH
levels. Tritiated water entered foliage at similar rates
at all pH values tested. In addition to differences in
uptake due to acidity, incorporation rates of various ions
may differ markedly. Water penetrated leaves much faster

than $^{86}Rb^+$. In addition water molecules entered foliage
about one thousand times faster than sulfate ions.[28]
These results suggest that absorption of materials in
water on leaf surfaces is a selective process that may be
affected by solution acidity, cellular metabolism, and
possibly other factors. However, individual ions appear
to have their own specific rates of entry which may be
affected by both intrinsic and extrinsic factors. These
data alone would suggest that anions may penetrate foliage
faster than cations at low pH. Such anion-cation generalizations do not appear to be warranted since ^{63}Ni, ^{65}Zn,
and ^{36}Cl penetrated leaves faster at lower pH levels
(2.7-3.0) than at pH 5.7. In general, penetration increased
as time of exposure increased. ^{65}Zn was incorporated into
foliage more rapidly than the other two isotopes used.[63]

It is possible that nitrate, sulfate and other
elements in acidic rain might stimulate plant productivity.
However, little information about inputs of ions from rain
per se into foliage is available. There has been a great
deal of interest in the application of nutrients by foliar
sprays to improve plant productivity, but it must be
remembered that surfactants are used in such experiments
to increase foliar penetration. Since surfactants are not
present in natural rain, the results of these agricultural
experiments are not directly comparable to those obtained
under natural conditions. They can, however, be used to
provide some perspective on possible contributions. No
consistent yield responses were obtained from soybean
plants receiving four foliar spray applications, each
supplying 80 kg ha^{-1} nitrogen, 8 kg ha^{-1} phosphorus, 24 kg
ha^{-1} potassium and 4 kg ha^{-1} sulfur when applied during
the optimal stages of plant growth for increasing seed
yield by foliar fertilization (i.e., total of 320 kg ha^{-1}
of nitrogen (as urea) and 16 kg ha^{-1} of sulfur).[68,69]
Throughout the soybean growing season of 1981 at Upton,
New York, ambient rainfall supplied 1.80 kg nitrogen and
2.75 kg sulfur ha^{-1}. In the soybean ambient rainfall
exclusion experiment described above, a total of 18.1 kg
nitrogen ha^{-1} as nitrate and 102 kg sulfur ha^{-1} as sulfate
were applied in simulated rainfalls at pH 2.7. These large
amounts of nitrate and sulfate did not counteract the
negative effects of rainfall acidity on soybean yields.

No consistent increases in yields of corn[69] and rice[70]
have been obtained. In addition, two experiments performed

with foliar fertilization of Zea mays under the rain exclusion facility[71] demonstrated no beneficial effect. During
1 year, a 6.4% reduction in seed yield occurred when plants
were exposed to 22.5, 4.5, 4.5 and 2.3 kg ha^{-1} nitrogen,
phosphorus, potassium and sulfur respectively, commencing
2 weeks before silking. It would seem reasonable to
conclude that plant productivity is not improved through
foliar absorption of nitrogen and sulfur from current
ambient rainfall and/or simulated rainfall under conditions
similar to ambient. Similarly, it is believed that the
amounts of nitrogen and sulfur added to soils by rainfall
are insignificant compared with amounts added through
routine fertilization.[1,28] This is supported by ^{15}N
tracer studies showing that the amount of nitrogen incorporated into the foliage of bean plants from simulated rain
solutions is insignificant compared with nitrogen inputs
from other sources.[72]

Changes in Leaf Cell Permeability of Leaves Exposed to Rainfall Acidity

Since H$^+$ ions may penetrate leaves, changes in leaf
cell permeability after exposure were tested.[28] Isotopes
were added to nutrient solutions after a single 20 min
rainfall at various pH levels. Twenty-four hours after
the first rainfall a second 10-min rain was applied. After
the foliage dried, leaf discs were placed in 0.5 mM CaSO$_4$
solutions for 120 min. For ^{86}Rb$^+$, ^{3}H$^+$, and ^{35}SO$_4^=$ there
was not significant differences in leaching rates among the
acidity treatments.

Cell permeability to ^{63}Ni was lowest at a simulated
rainfall of pH 3.0. The low permeability observed at pH
3.0 could not be rationalized with high permeability at
pH 3.4 and intermediate permeability at both pH 5.7 and
2.7. Between 45 and 68% of the ^{63}Ni in the foliage was
permeable during the 120-min time period. Rainfall acidity
had no effect on cell permeability to ^{65}Zn. Moreover,
permeability was constant throughout incubation and ranged
from 47 to 75% among the pH levels tested. Cell permeability
of ^{36}Cl from foliage was strongly influenced by simulated
rainfall acidity. Permeability at pH 2.7 was about 3 times
higher than that at pH 5.7. Mean cell permeability to ^{36}Cl
was about 70%. These results taken together suggest that
most elements tested did not show any differences in cell
permeability. However, the fact that two of the six

elements tested showed a significant acidity effect suggests that the acidity of the test solutions in simulated rainfall can alter membrane permeability.

Removal of Materials From Foliage by Rain

Materials can be lost from leaf surfaces by guttation and secretion, or by leaching by rain, dew, and mists. Marked effects can result from such losses. For example, Dalbro (reported in Ref. 30) reported marked losses of potassium (25-30 kg/ha), sodium (9 kg/ha), and calcium (10.5 kg/ha, annually) from apple orchards. Most literature confirms that the highest leaching rates occur from older leaves, i.e., those that have experienced the most weathering.

Since plants produce substances on their surfaces, it is of interest to determine if the removal of these substances is sensitive to acidity in precipitation. Likewise, substances within leaves may be released after exposure to rainfalls. Moreover, the nutrient levels in harvested portions of crops may affect the quality of foodstuffs. It is conceivable that acid precipitation could sufficiently influence nutrient leaching from plant surfaces so as to alter crop quality. Wood and Bormann[73] demonstrated that K^+, Ca^{2+}, and Mg^{2+} were leached from pinto bean leaves more rapidly at pH values of 3.0 and 3.3 than at pH 3.3 from foliage of sugar maple. Leaching rates of K^+ and Mg^{2+} were higher at pH 3.0 than at pH 4.0. In tobacco leaves, Ca^{2+} leached faster from foliage exposed to simulated rainfalls of pH 3.0 than from foliage exposed to pH 6.7.[74] In pinto beans exposed to a 20-min daily treatment of simulated rain for five days, more calcium, nitrate, and sulfate were leached from foliage at pH levels of 2.7, 2.9, and 3.1 than from plants exposed to pH 5.7.[28] In contrast, the amount of potassium leached was greater from leaves exposed to pH 5.7 than from leaves exposed to pH levels between 3.4 and 2.9. The amounts of ammonium, magnesium, and zinc leached were the same at all pH levels tested. Three radioisotopes were detected in leachates from leaflets of Phaseolus vulgaris that were exposed to one 20-min rainfall 24 hours after radioisotope addition. Although the quantities of ^3H_2O, $^{86}Rb^+$, $^{35}SO_4^=$ were less than 1% of the radioisotopes in leaflets, these results demonstrate that recently absorbed ions may be leached by rainfall.

In later experiments, tests were made using three additional radioactive elements.[63] Zinc was studied because it is an essential element for plants and is present in ambient deposition. Nickel was examined because it is a non-essential metal. Chloride was selected because it is an anion and comparisons with sulfate are germane. Leaching of recently absorbed ^{63}Ni was <2% while values for ^{65}Zn and ^{36}Cl ranged between 3 and 6%. With ^{63}Ni, leaching was greater at pH 5.7 than at low pH, while the opposite situation occurred with ^{36}Cl. Leaching of ^{65}Zn was independent of pH. These results substantiated earlier studies of Evans et al.[28]

Hindawi et al.[75] demonstrated lower amounts of nitrogen, calcium, magnesium and phosphorus in foliage of P. vulgaris L. exposed to simulated acid mist. In general, the percentage decrease of these four elements was greater than the percentage decreases in plant dry weights among the pH levels examined. Hindawi et al.[75] did not detect any differences in potassium contents in foliage among treatments (pH range 5.5 - 2.0). A marked increase in foliar sulfur was present in plants exposed to elevated levels of sulfate (and acidity) in the mists.

Cole and Johnson[76] measured the pH and conductivity (the latter a measure of total cations and anions) of ambient precipitation and throughfall during two rainfalls in 1973. During the latter portion of the first rainfall, the pH of throughfall (4.0) was about 1 pH unit below that of ambient precipitation (5.0). Conductivities of the two solutions were similar at about 10 μmho cm^{-1}. During the second rainfall, the pH of ambient precipitation (3.5) was almost 2 units below that of the throughfal (5.3). Over an entire annual cycle, the amount of sulfur in throughfall was greater than that in ambient precipitation. Cole and Johnson[76] concluded that the differences in sulfur content observed were due to foliar leaching. Moreover, van Breeman et al.[77] demonstrated that ammonium, nitrate, and sulfate volume-weighed concentrations were higher in throughfall and stemflow than ambient rain from forests at two locations in the Netherlands. These results suggest that there is a net leaching of nitrogen and sulfur from these species when exposed to rainfalls of pH 4.29 and 4.51. However, washoff of dry deposition may have contributed to higher sulfate concentrations in through-fall.

These results coupled with results of other investigators show that there is no general relationship between
the pH of incident rainfall and that of throughfall among
various plant species. To further complicate the
situation, little is known about the chemistry and
relative acidifying effects of dry deposition. Methods
to measure dry deposition are not available. Although
results of such studies attempt to understand how acidic
precipitation affects foliage, limited conclusions can be
drawn about the overall effects of altered ion mobility
and leachability when periods of one or more growth
seasons are considered. As a result, relationships
between droplet pH, ion mobility, throughfall chemistry
and infections by plant pathogens remain unknown under
natural field or forest conditions.

Our results, coupled with results of other investigators, have shown that acidic precipitation affects the
leachability and mobility of elements in or on foliage.
These alterations in ion mobility and nutrient status may
occur in the absence of visible leaf injury. Rainfall
acidity can change the chemistry of leaf surfaces and
cells within leaves. However, there are no data to
document that there is a significant influx of materials
in ambient rain that benefit plants or removal by
leaching that would limit plant productivity. This lack
of documentation might be due to the fact that our knowledge of nutrient cycling, transport, and accumulation are
insufficient for such an overall evaluation at present.
Only well-designed experiments will provide adequate
information to determine if these physiological changes
may influence other important functions such as growth,
development and yields of commercial crops or natural
terrestrial ecosystems.

SIGNIFICANCE OF CHANGES IN MINERAL NUTRITION WITH PLANT
PRODUCTIVITY BY ACIDIC DEPOSITION

A large body of knowledge is available to evaluate
how changes in nutrient elements affect plant productivity.[78] Unfortunately, unless a relatively complete
analysis of a variety of nutrients and an evaluation of
plant growth and development is available, the significance of acidic deposition on plant productivity by
changes in mineral nutrition and metabolism is now

sufficiently known so that cause and effect relationships can be recognized. At the present time, firm relationships between acidic deposition and alterations in mineral or carbohydrate metabolism, coupled with changes in productivity of field-grown soybeans cannot be established. Experimentation in this area is warranted.

There is speculation that nitrogen inputs from precipitation, including cloud water, have a significant effect on high elevation red spruce (Picea rubens Sarg.) trees in northeastern United States.[79] Present information does not exclude this possibility but further research is needed to establish a cuase and effect relationship. Current information suggests that the nitrogen content of cloud water is much higher (3 to 10 times) than rain or snow[80] and that cloud occlusion is frequent at high elevations. It is speculated that the nitrogen from suggested deposition sources could alter red spruce tree metabolism enough to affect winter/frost hardiness. This avenue of research is also warranted.

TESTABLE HYPOTHESES TO DETERMINE IF ACIDIC DEPOSITION COULD SIGNIFICANTLY ALTER PLANT PRODUCTIVITY THROUGH CHANGES IN NUTRIENT AND CARBOHYDRATE METABOLISM

Research has shown that acidic deposition negatively affects field-grown soybeans. Research has also shown that high elevation red spruce is experiencing growth reductions and that atmospheric depositions may play a role. These two research areas would appear to have the greatest potential for understanding how atmospheric deposition impacts terrestrial vegetation. It is highly recommended that research be conducted under conditions that most closely mimick field conditions. This is necessary because the effects of acidic deposition on soybeans differ under field conditions compared with controlled-environment conditions, and the growth and survival of red spruce is surely controlled by temperature, length of growing season, light intensity and duration, and other factors.

An overall hypothesis for effects of acidic deposition on field-grown soybeans could be that exposure to acidic deposition decreases soybean seed yields through significant alterations in mineral metabolism through changes in cell permeability and/or removal of materials from foliage.

Testable hypotheses for this overall hypothesis could be:

(1) Acidic deposition significantly alters nitrogen and carbon metabolism of field-grown soybeans. Concentrations of important nitrogen compounds such as proteins, hydroxylamine, nitrogen, and inorganic amides should be determined in relation to plant development and growth stages.

(2) Acidic deposition significantly alters mineral nutrition of calcium, potassium, phosphorus, sulfur, and possibly other nutrients of field-grown soybeans. These studies would require elemental analyses in relation to a variety of plant developmental and growth stages.

An overall hypothesis for effects of atmospheric deposition on high elevation red spruce in the northeastern United States could be that exposure of red spruce to acidic deposition significantly alters frost/ winter hardiness of young twigs and needles. Data on the following could provide pertinent information.

(1) Atmospheric deposition, specifically nitrogen deposition from cloud water, can alter normal growth, periderm formation, and winter hardiness. The excess nitrogen can provide for a surplus of metabolically active proteins that could result in excessive vegetative growth with inaccurate/delayed perception of environmental cues for winter hardiness protection.

(2) Atmospheric deposition could alter cuticular features that can provide excessive winter desiccation injury of young shoots and needles. Shoot and needle cuticular and other surface features could be analyzed under high elevation forest conditions in which various deposition regimes are imposed.

REFERENCES

1. EVANS, L.S., G.R. HENDREY, G.J. STENSLAND, D.W.
   JOHNSON, A.J. FRANCIS. 1981. Considerations of
   an air quality standard for acidic precipitation.
   Water Air Soil Pollut. 16: 469-509.
2. LIKENS, G.E., F.H. BORMANN, N.M. JOHNSON. 1972.
   Acid rain. Environment 14: 33-44.
3. NORDO, J. 1976. Long range transport of air
   pollutants in Europe and acid precipitation in
   Norway. Water Air Soil Pollut. 6: 199-227.
4. ODEN, S. 1976. The acidity problem - An outline
   of concepts. Water Air Soil Pollut. 6: 137-166.
5. JACOBSON, J.S., L.I. HELLER, P. VAN LEUKEN. 1976.
   Acidic precipitation at a site within the northeastern conurbation. Water Air Soil Pollut. 6:
   339-349.
6. YUE, G.K., V.A. MOHNEN, C.S. KIANG. 1976. A
   mechanism for hydrochloric acid production in
   cloud. Water Air Soil Pollut. 6: 277-294.
7. GALLOWAY, J.N., D.M. WHELPDALE. 1980. An atmospheric sulfur budget for eastern North America.
   Atmos. Environ. 14: 409-417.
8. STENSLAND, G.J. 1983. Wet deposition network data
   with applications to selected problems. Chapter
   4 in Volume 1, A.P. Altschuller (ed.), Atmospheric sciences (the acidic deposition phenomenon
   and its effects). EPA-600/8-83-016A.
9. EVANS, L.S., T.M. CURRY. 1979. Differential
   responses of plant foliage to simulated acid rain.
   Am. J. Bot. 66: 953-962.
10. EVANS, L.S., K.F. LEWIN, C.A. CONWAY, M.J. PATTI.
    1981. Seed yields (quantity and quality) of
    field-grown soybeans exposed to simulated acidic
    rain. New Phytol. 89: 459-509.
11. EVANS, L.S., K.F. LEWIN, M.J. PATTI, E.A. CUNNINGHAM.
    1983. Productivity of field-grown soybeans
    exposed to simulated rain. New Phytol. 93:
    377-388.
12. EVANS, L.S., K.F. LEWIN, M.J. PATTI. 1984. Effects
    of simulated acidic rain on yields of field-grown
    soybeans. New Phytol. 96: 207-213.
13. EVANS, L.S., K.F. LEWIN, K.A. SANTUCCI, M.J. PATTI.
    1985. Effects of frequency and duration of
    simulated acidic rainfalls on soybean yields.
    New Phytol. 100: 199-209.

14. EVANS, L.S., K.F. LEWIN, E.M. OWEN, K.A. SANTUCCI. 1986. Comparison of several cultivars of field-grown soybeans exposed to simulated acidic rainfalls. New Phytol. 102: 409-417.

15. BANWART, W.L., J.J. HASSETT, B.L. VASALIS. 1984. Acid rain and its effect on corn and soybean yields. Proceedings of the Illinois Fertilizer and Chemical Dealers' Conference, pp. 19-21.

16. IRVING, P.M., J.E. MILLER. 1981. Productivity of field-grown soybeans exposed to acid rain and sulfur dioxide alone and in combination. J. Environ. Qual. 10: 473-478.

17. TROIANO, J., L. COLAVITO, L. HELLER, D.C. McCUNE, J.S. JACOBSON. 1983. Effects of acidity of simulated rain and its joint action with ambient ozone on measures of biomass and yield in soybean. Environ. Exp. Bot. 23: 113-119.

18. EVANS, L.S., L. DIMITRIADIS, D.A. HINKLEY. 1984. Seed protein quantities of field-grown soybeans exposed to simulated acidic rain. New Phytol. 97: 71-76.

19. EVANS, L.S., M.J. SARRANTONIO, E.M. OWEN. 1986. Protein contents of seed yields of field-grown soybeans exposed to simulated acidic rain: Assessment of the sensitivities of four cultivars and effects of duration of simulated rainfall. New Phytol. 103: 689-693.

20. GALLOWAY, J.N., G.E. LIKENS. 1976. Calibration of collection procedures for the determination and precipitation chemistry. Water Air Soil Pollut. 6: 241-258.

21. LIKENS, G.E., F.H. BORMANN, J.S. EATON, R.S. PIERCE, N.M. JOHNSON. 1976. Hydrogen input to the Hubbard Brook Experiment Forest, New Hampshire, during the last decade. Water Air Soil Pollut. 6: 435-445.

22. SEMB, A. 1976. Measurement of acid precipitation in Norway. Water Air Soil Pollut. 6: 231-240.

23. KELLOGG, W.W., R.D. CADLE, E.R. ALLEN, A.L. LAZRUS, E.A. MARTELL. 1972. The sulfur cycle. Science 175: 587-596.

24. FRIEND, J.P. 1979. Sulfur compounds and their distributions. Conference on aerosols: Anthropogenic and natural sources and transport. N.Y. Acad. Sci. January 9-12, 1979, New York City.

25. National Atmospheric Deposition Program. 1978 and
    1979. NADP data reports. Vol. II (1-3).
    (Available from NADP Coordinator's Office,
    Natural Resource Ecology Laboratory, CSU, Fort
    Collins, Colorado.)
26. PACK, D.H. 1980. Precipitation chemistry patterns:
    A two-network data set. Science 208: 1143-1145.
27. RAYNOR, G.S., J.V. HAYES. 1978. Experimental data
    from analysis of sequential precipitation samples
    at Brookhaven National Laboratory. BNL Report
    No. 50826.
28. EVANS, L.S., T.M. CURRY, K.F. LEWIN. 1981.
    Responses of leaves Phaseolus vulgaris to
    simulated acidic rain. New Phytol. 88: 403-420.
29. EVANS, L.W., G.S. RAYNOR, D.M.A. JONES. 1984.
    Frequency distributions for durations and volumes
    of rainfalls in the eastern United States in
    relation to acidic precipitation. Water Air Soil
    Pollut. 23: 187-195.
30. MARTIN, J.T., B.E. JUNIPER. 1970. The cuticles of
    plnats. St. Martin's Press, New York, 347 p.
31. JUNIPER, B.E. 1960. Studies on structure in
    relation to phytotoxicity. Ph.D. Dissertation,
    University of Oxford, Great Britain.
32. HALL, D.M., A.I. MATUS, J.A. LAMBERTON, H.N. BARBER.
    1965. Infra-specific variation in wax on leaf
    surfaces. Aust. J. Biol. Sci. 18: 323-332.
33. LINSKENS, H.F. 1950. Quantitative Bestimmung der
    Benetzbarkeit von Blattoberflächen. Planta 38:
    591-600.
34. FOGG, G.E. 1948. Adhesion of water to external
    surfaces of leaves. Discuss. Faraday Soc. 3:
    162-169.
35. CRAFTS, A.S. 1961. The chemistry and mode of action
    of herbicides. Interscience Publishers, New York,
    269 p.
36. SCHNEPF, E. 1965. Licht- und elektronenmikros-
    kopische Beobactungen an der Trichom-Hydathoden
    von Cicer arietinum. Z. Pflanzenphysiol. 53:
    245-254.
37. HABERLANDT. G.F.J. 1914. Physiological plant
    anatomy. Macmillan, London, 777 p.
38. KONAR, R.N., H.F. LINSKENS. 1966. The morphology
    and anatomy of the stigma of Petunia hybrida.
    Planta 71: 356-371.

39. ADAM, N.K.   1948. Principles of penetration of
     liquids into solids. Discuss. Faraday Soc. 3:
     5-11.

40. GUSTAFSON, F.G.   1956. Absorption of Co^{60} by leaves
     of young plants and its translocation through the
     plant. Am. J. Bot. 43: 157-160.

41. GUSTAFSON, F.G.   1957. Comparative absorption of
     cobalt-60 by upper and lower epidermis of leaves.
     Plant Physiol. 32: 141-142.

42. SARGENT, J.A., G.E. BLACKMAN.   1962. Studies on
     foliar penetration. 1. Factors controlling the
     entry of 2,4-dichloroacetic acid. J. Exp. Bot.
     13: 348-368.

43. SCHONHERR, J., M.J. BUKOVAC.   1972. Penetration
     of stomata by liquids: Dependence on surface
     tension, wettability, and stomatal morphology.
     Plant Physiol. 49: 813-819.

44. NORMAN, A.G., C.E. MINARIK, R.L. WEINTRAUB.   1950.
     Herbicides. Annu. Rev. Plant Physiol. 1: 141-168.

45. ROBERTS, E.A., M.D. SOUTHWICK, D.H. PALMITER.   1948.
     A microchemical examination of McIntosh apple
     leaves showing relationship of cell wall constit-
     uents to penetration of spray solutions. Plant
     Physiol. 23: 557-559.

46. SCHONHERR, J.   1976. Water permeability of isolated
     cuticular membranes: The effect of pH and cations
     on diffusion, hydrodynamic permeability and size
     of polar pores in the cutin matrix. Planta 128:
     113-126.

47. SCHONHERR, J., H.W. SCHMIDT.   1979. Water perme-
     ability of plant cuticles: Dependence of
     permeability coefficients of cuticular transpi-
     ration on vapor pressure saturation deficit.
     Planta 144: 391-400.

48. ORGELL, W.H. 1957. Sorption properties of plant
     cuticle. Proc. Iowa Acad. Sci. 64: 189-197.

49. McFARLANE, J.C., W.L. BERRY.   1974. Cation penetra-
     tion through isolated cuticles. Plant Physiol.
     53: 723-727.

50. PRASAD, R., G.E. BLACKMAN.   1962. Factors affecting
     the uptake of 2,2-dichloropropionic acid by roots
     and fronds of Lemna minor. Plant Physiol. 37:
     xiii (suppl.).

51. BENNETT, S.H., W.D. THOMAS.   1954. The absorption,
     translocation and breakdown of schraden applied

to leaves, using ^{32}P labeled material. Ann. Appl. Biol. 41: 484-500.

52. JYUNG, W.H., S.H. WITTWER. 1964. Foliar absorption - An active uptake process. Am. J. Bot. 51: 437-444.

53. DYBING, C.D., H.B. CURRIER. 1961. Foliar penetration of chemicals. Plant Physiol. 36: 169-174.

54. LEONARD, O.A. 1958. Studies on the absorption and translocation of 2,4-D in bean plants. Hilgardia 28: 115-160.

55. EVANS, L.S., N.F. GMUR, J.J. KELSCH. 1977. Perturbations of upper leaf surface structures by simulated acid rain. Environ. Exp. Bot. 17: 145-149.

56. EVANS, L.S., N.F. GMUR, F. DaCOSTA. 1977. Leaf surface and histological perturbations of leaves of Phaseolus vulgaris and Helianthus annuus after exposure to simulated acid rain. Am. J. Bot. 64: 903-913.

57. EVANS, L.S., N.F. GMUR, F. DaCOSTA. 1978. Foliar response fo six clones of hybrid poplar to simulated acid rain. Phytopathology 68: 847-856.

58. PAPAROZZI, E.T. 1981. The effects of simulated acid precipitation on leaves of Betula alleghaniensis Britt. and Phaseolus vulgaris cv. red kidney. Ph.D. Thesis, Cornell University, Ithaca, New York.

59. ADAMS, C.M. 1982. The response of Artemisia tilesii to simulated acid precipitation. M.S. Thesis, University of Toronto, Ontario.

60. SHRINER, D.S. 1974. Effects of simulated rain acidified with sulfuric acid on host-parasite interactions. Ph.D. Thesis, North Carolina State University, Raleigh, North Carolina.

61. HOLLOWAY, P.J., E.A. BAKER. 1968. Isolation of plant cuticles with zinc chloride-hydrochloric acid solution. Plant Physiol. 43: 1878-1879.

62. UPHOF, J.D. Th. 1962. Plant hairs. Gebruder Borntraeger, Berlin-Nokolassee.

63. EVANS, L.S., K.A. SANTUCCI, M.J. PATTI. 1985. Interactions of simulated rain solutions and leaves of Phaseolus vulgaris L. Environ. Exp. Bot. 25: 31-41.

64. GALLOWAY, J.N., J.D. THORNTON, S.A. NORTON, H.L. VOLCHOK AND R.A. McLEAN. 1982. Trace metals

in atmospheric deposition: A review and assessment. Atmos. Environ. 16: 1677-1700.

65. LINDBERG, S.E. 1982. Factors influencing trace
metal sulfate, and hydrogen ion concentrations in
rain. Atmos. Environ. 16: 1701-1709.

66. LINDBERG, S.E., R.R. TURNER, D.S. SHRINER, D.D. HUFF.
1981. Atmospheric deposition of heavy metals and
their interaction with acid precipitation in a
North American deciduous forest. In Third
International Conference on Heavy Metals in the
Environment. (S.E. Lindberg, R.R. Turner, eds.),
Oak Ridge National Laboratory, Oak Ridge,
Tennessee, pp. 306-309.

67. REUTHER, R., R.F. WRIGHT, U. FÖRSTNER. 1981.
In S.E. Lindberg, R.R. Turner, eds., ibid.
Reference 66., Oak Ridge National Laboratory,
Oak Ridge, Tennessee, pp. 318-332.

68. GARCIA, R.L., J.J. HANWAY. 1976. Foliar fertilization of soybeans during the seed-filling period.
Agron. J. 68: 653-657.

69. NEUMANN, P.M., Y. EHRENREICH, Z. GOLAB. 1981. Foliar
fertilizer damage to corn leaves: Relation to
cuticular penetration. Agron. J. 73: 979-982.

70. THOM, W.O., T.C. MILLER, D.H. BORMAN. 1981. Foliar
fertilization of rice after midseason. Agron. J.
73: 411-414.

71. HARDER, H.J., R.E. CARLSON, R.H. SHAW. 1982. Leaf
photosynthetic response to foliar fertilizer
applied to corn plants during grain fill. Agron.
J. 74: 759-761.

72. EVANS, L.S., D.C. CANADA, K.A. SANTUCCI. 1986.
Foliar uptake of ^{15}N from rain. Environ. Exp.
Bot. 26: 143-146.

73. WOOD, J., F.H. BORMANN. 1975. Increase in foliar
leaching caused by acidification of an artificial
mist. Ambio 4: 169-171.

74. FAIRFAX, F.A., N.W. LEPP. 1975. Effect of simulated "acid rain" on cation loss from leaves.
Nature 225: 324-325.

75. HINDAWI, I., J.A. REA, W.L. GRIFFIS. 1979.
Response of bush bean exposed to acid mist.
70th Annu. Meeting Joint Air Pollution Control
Assoc. Abstract 77, 30.4.

76. COLE, D.W., D.W. JOHNSON. 1977. Atmospheric sulfate
additions and cation leaching in a Douglas-fir
ecosystem. Water Resources Res. 13: 313-317.

77. vanBREEMEN, N., P.A. BURROUGH, E.J. VELHORST, H.F.
     van DOBBEN, T. de WIT, T.B. RIDDER, H.F.
     REIJNDERS. 1982. Soil acidification from
     atmospheric ammonium sulfate in forest canopy
     throughfall. Nature 299: 548-550.
78. HEWITT, E.J., T.A. SMITH, eds. 1974. Plant mineral
     nutrition. English Universities Press, London,
     298 p.
79. EVANS, L.S. 1986. Proposed mechanisms of initial
     injury causing apical dieback in red spruce at
     high elevation in eastern North America. Can. J.
     For. Res., in press.
80. SCHERBATSKOY, T., R.M. KLEIN. 1983. Response of
     spruce and birch foliage to leaching by acidic
     mists. J. Environ. Qual. 12: 189-195.

Chapter Nine

BIOAVAILABILITY OF HEAVY METALS IN SLUDGE-AMENDED SOILS
TEN YEARS AFTER TREATMENT

CHARLES L. MULCHI, PAUL F. BELL,
CHARLES ADAMU AND JOSEPH R. HECKMAN

Department of Agronomy
University of Maryland
College Park, Maryland 20742

INTRODUCTION

The application of municipal sewage sludge to
agricultural land has become an established practice in
the United States, especially in areas adjacent to major
metropolitan centers.[1] The U.S. Environmental Protection
Agency (EPA) established guidelines for regulating the
application of sludge materials on farmlands in the
1970s.[2] However, several agricultural experiment stations
in the highly urbanized northeastern states, including
Maryland, have established their own guidelines for the
disposal of sewage sludge on crop lands.[1-3] The primary
factors which limit the quantities of sludge that may be
applied to an area are the amounts of heavy metals
present in a given sludge and the nutrient requirements of
the crops being grown.[1,2] The N and P requirements of the
crop generally limit the quantity of sludge that can be
applied in a given year and the amount of heavy metals

235

applied to soils are commonly used to establish lifetime
loading limits.[1-8,10,13,14]

The amounts and ratios of heavy metals in sludge
materials will vary greatly and usually reflect the
residential or industrial character of the city or
district supplying the raw sewage. Sludges from small
towns and cities with high residential and low industrial
inputs into the local wastewater treatment facilities will
typically exhibit much lower quantities of heavy metals
than is found in products from major industrial centers
such as Chicago, Milwaukee or Baltimore.[2,7] Also, there
is an array of wastewater treatment processes, such as
liming, that may alter the chemical and physical
properties of the sludge materials.[2,9,10]

Most of the guidelines and recommendations concerning
land application of municipal sludges were based on crop
responses determined within a short time (i.e., 1 to 4
years) following sludge application with major concern
for Cd uptake by plants.[1-4,8,11-14] Chemical reactions
in the soil immediately following application of sludge
may differ greatly from those found at a later date.
Municipal sludge is typically composed of a mixture of
solids, liquids, soluble salts or ash and organic matter.[10]
The levels of soluble salts and organic matter in the
amended soil are much higher immediately following
application than would be found a decade later. Also,
rates of organic matter decomposition would be expected
to decline as the sludge amendments reach equilibrium
with the soil.[1,3,10]

The products of organic matter decomposition in soil
includes lignin and NH_4^+, both of which can facilitate the
uptake of metals by plant roots.[1,3,5,6,9,10,12] The
metals form metal-ligand complexes in the soil media which
can facilitate metal diffusion and uptake by plants.
Increases in the hydrogen ion activity associated with the
metabolism of NH_4^+ being released during organic matter
decomposition may also enhance the availability for most
heavy metals in sludge amended soil.[3-5,9] Therefore,
metal uptake responses immediately following sludge
application would likely be higher than after the organic
matter has reached equilibrium in the soil. However, the
number of case studies on long term changes in soil
chemical processes in sludge amended soils are limited.[8]

The manner in which soil is managed following sludge application may also influence the long-term bioavailability of metals. To avoid problems with metal toxicities by metals such as Zn and crop contamination by metals such as Cd, it is commonly recommended that sludge amended soils be maintained at a pH of 6.5.[1,2,4,9,10,12,14] As the amount of sludge amended soil increases, the level of site monitoring to insure that the guidelines are being properly followed is likely to decrease due to limited resources. The problem is further complicated by the absence of readily available land records for sludge amended areas. During sales or leasing of farmlands, the new owners or operators may not be properly informed of the sludge amended areas that demand special attention in order to avoid problems with metal toxicities or contamination of farm products grown on these soils.[1,11-13]

The objectives of the studies summarized in this article were: (1) to acquire information on the long-term bioavailability of heavy metals in soils amended at various rates using sludge products from several municipal sources; (2) to compare metal accumulation responses for tobacco (Nicotiana tabacum L.) and soybenas (Glycine max Meri.) grown on sludge amended soils; and (3) to assess the influence of soil management or cropping system on the bioavailability of metals in sludge amended soils.

MATERIALS AND METHODS

Research Sites

The information supplied in this manuscript resulted from studies conducted during 1983-1985 at three research sites located in Maryland that were established during the period 1972-1978. Sites I and II were located at the University of Maryland's (UMD) Plant Research Farm near Fairland, Maryland which was recently sold to developers. Site III is located on the USDA's Hayden Farm at the Beltsville Agricultural Research Center (BARC) near Beltsville, Maryland.

Site I. Dewatered, digested sludge from the Blue Plains wastewater treatment facility in Washington, D.C. was applied in strips 7.3 x 42 m at rates equal to 0, 56, 112 and 224 Mg ha^{-1} (dry weight) in 1972 (Table 1). The

Table 1. Summary of metal concentrations in the sewage sludges applied to the UMD and USDA–BARC research sites.

Research[+] Site	Sludge Source	Year Applied	Sludge Characteristics	Fe (x10⁴)	Metals (mg kg^{-1})					
					Zn	Cu	Mn	Pb	Ni	Cd
I. UMD	Blue Plains Washington, D.C.	1972	Dewatered Digested	2.5	1,302	573	296	276	45	13
II. UMD	Back River Baltimore, MD.	1975	Anaerobically Digested	--	4,400	2,200	--	140	170	16
III. USDA–BARC	Piscataway PG Co., MD.	1976	Limed Digested	2.5	639	259	722	217	15	5.9
III. USDA–BARC	Piscataway PG Co., MD.	1976	Limed Raw	2.5	599	277	598	215	17	4.9
III. USDA–BARC	Annapolis, MD.	1976	Heat Treated	8.3	1,329	404	854	360	37	13.4
III. USDA–BARC	Blue Plains Washington, D.C.	1976	Limed Compost	4.1	731	274	719	272	201	7.2
III[‡]. USDA–BARC	Nu-Earth Chicago, IL.	1978	Digested	2.5	4,140	1,160	302	865	590	210
Maximum		Domestic	Sludge[††]	--	2,500	1,000	--	100	200	25

[+]The UMD research site was located on the Plant Research Farm near Fairland, MD. (2)
The USDA–BARC research site was located on the Hayden Farm near Beltsville, Md. (10)

[‡]Information on this treatment was included because it is representative of a high metal sludge of industrial origin which is commercially available.

[††]From Chaney and Giordano (1977).

soil at this site is classified as Beltsville silt loam
(fine, loamy, mixed, mesic typic Fragiudults). The soil
appeared uniform, flat and well drained. The metal
contents of the sludge applied to Site I (Table 1) are
consistent with values found for sludges from non-
industrial sources.[5] The amounts of metals applied at
the highest sludge application rate (224 Mg ha^{-1} were
291, 128, 66, 61, 10 and 2.1 kg ha^{-1} for Zn, Cu, Mn, Pb,
Ni and Cd respectively. These values are below the
lifetime loading limits set by the EPA which equal 500,
250, 1000, 250, and 5 kg ha^{-1} for Zn, Cu, Pb, Ni and
Cd.[1,5]

Observational units (i.e., plots) were established by
dividing each strip of 42 m into 6 segments 7 m in length.
Three of the adjoining segments were designated as site A,
the remaining three as site B.

Previous investigators were interested in examining
the interaction of supplemental fertility and sludge
treatments. This was accomplished by further subdividing
the plots into subplots 3.6 x 7.0 m and randomly applying
supplemental P and K to half of the subplots in site A
and supplemental N, P and K in site B. Sites A and B
were managed continuously for soybeans (Glycine max, Meri.)
and corn (Zea mays L.), respectively, from 1972 through
1981 with the same field design being followed over the
10 growing seasons.[3] Prior to planting sites A and B to
tobacco in 1983 and 1984, both sites were uniformly
fertilized with N, P and K at rates equal to 68, 134
and 202 kg ha^{-1}, respectively, using a commercial fertilizer labeled for tobacco.[3]

In an effort to accelerate changes in soil acidity
which would likely occur with long-term farm activities,
the fertility subplots were further subdivided into 1.8 x
7.0 m areas and elemental S was randomly applied to half
of the subunit areas in May, 1983. Because of the higher
buffering capacity anticipated for the soil which
received the higher sludge rates, areas treated with
sludge at 112 and 224 Mg ha^{-1} received the equivalence
of 900 kg ha^{-1} S compared to 720 kg ha^{-1} for the 0 and
56 Mg ha^{-1} sludge treatments. The desired effect from
the S additions was to induce pH changes of the order of
0.5 to 1.0 units. The S was incorporated in the soil
using a tractor P.T.O. power tiller.

Seedlings of Maryland tobacco cultivar "Md 609" were
grown in beds at the University of Maryland's Tobacco
Experimental Farm near Upper Marlboro, Maryland and
transplanted to the field in early June when plants
averaged 12 cm in height. The rows were 0.9 m apart with
plants spaced 0.6 m within the row. The observation rows
were in the center of the S treated areas and were bordered
by single rows of plants in all cases. Supplemental N
was applied three weeks after transplanting as NH_4NO_3 at a
rate of 33 kg ha^{-1}. Recommended cultural practices for the
growth of Maryland tobacco were followed. Due to
inadequate rainfall during July, 1983, 2.5 cm of water
was applied to both sites as sprinkler irrigation. Rainfall amounts during 1984 were adequate for normal growth.
Ten plants per plot were harvested in late August and
placed in barns for air curing. Upon curing the leaves
were removed from the stalks and sorted into lots having
similar physical properties. Composite leaf samples were
collected on a weighed basis, ground in a stainless steel
mill and stored in glass containers[3] for later chemical
analysis.

Site II. Site II was established by Dr. Morris
Decker at the Plant Research Farm of UMD in 1975 using
a high metal, digested sludge from the Back River wastewater treatment facility in Baltimore, Maryland (Table 1).
The soil at this location is classified as a Sassafras
sandy loam (fine loamy, siliceous, mesic Typic Hapludults).
This site contained 12 replicates of sludge treatments 0,
56 and 112 Mg ha^{-1}. Soybeans, cultivar "Clark", were hill
planted using a 18 x 18 cm grid in plots 1.5 x 1.8 m.
The plants were grown to pod stage then composite samples
consisting of 10 whole plants (leaves, stems and pods)
were collected. After drying at 60°C, the samples were
ground and stored as previously described.[6]

Site III. Site III was established in 1976 by Dr.
Rufus Chaney on the Hayden Farm at USDA-BARC near
Beltsville, Maryland. The soil at the site is classified
as a Christiana fine sandy loam (clayey, kaolinitic, mesic
Typic Paleudults). The sludge research plots at USDA-BARC
were established to represent different sludge processing
technologies and metal concentrations (Table 1). Three
replicates of zero sludge controls at several soil pH
values were also available for comparisons among various
sludge treatments from several local sources and from

Chicago, Illinois (Nu-Earth). For ease of comparison, only data from the 0, 56, 112 and 224 Mg ha^{-1} sludge treatments established in 1976, and 50 and 100 Mg ha^{-1} Nu-Earth treatments established in 1978 are presented. Although the plots are 6.4 x 7.9 m in area, only half of the areas were used for any given crop.[6]

Soybeans were grown in areas 1.5 x 1.8 m in 1983 and 1984 to the pod stage of maturity then sampled as previously described.[6] Maryland tobacco was planted in June, 1984 and 1985 in three rows spaced 0.9 m apart with plants spaced 0.6 m apart. Also, N, P and K fertilizers were applied for the growth of tobacco as previously described; however, no supplemental fertilizers were applied to the soybeans. The tobacco was grown to maturity, harvested, cured and sampled as described for site I with 10 plants being harvested from the center row to serve as a composite sample.

Metal Analyses

Tobacco leaves. For determination of Zn, Cu, Mn, Fe, Pb, Ni and Cd, 1.0 g of ground tobacco was placed in a Kjeldahl digestion flask (100 ml) to which 10.0 ml of concentrated HNO_3 were added. After allowing predigestion with HNO_3 overnight, 5.0 ml of concentrated (70%) $HClO_4$ were added and the flask placed on a heater located within a ventilated hood equipped with a flue scrubber. Following digestion, the flask contents were filtered using Whatman No. 541 paper then transferred to a volumetric flask and diluted to 25 ml with distilled deionized water. Metal analyses were performed using a Perkin-Elmer atomic absorption spectrophotometer (Model 5000). Standards for each element were prepared from certified 1000 mg kg^{-1} element standard solutions. The standards for a given metal were prepared in a matrix containing fixed quantities of the six other metals. Deuterium background correction was used in the analysis for Cd, Pb and Ni.[3]

Soybean shoots. A 2.0 g subsample was dried and ashed at 500°C for 12 hours, dissolved in 4.0 ml concentrated HNO_3 and evaporated to dryness. The residue was then redissolved using 10 ml of 3N HCl, refluxed for 2 hours, filtered, and diluted to 25 ml with 0.1 N HCl.

Zinc, Cu, Mn, Cd and Ni were determined as described for
tobacco.[6]

Statistical Analyses

The results from the Site I tobacco studies were
analyzed annually as a split-split plot design with sludge
rates, fertility rates and S treatments as the main
plot, subplot and subsubplot variables, respectively.
Following the observations that fertility treatments were
not significant and that the variances for the corn and
soybean sites were homogeneous, the results were combined
in a single ANOVA. Since the results from 1984 were
observed to be consistent with the 1983 data, the results
from both years were combined. Sludge rates, S treatments,
crop history (i.e., corn vs. soybean management) and years
were found to be significant in a majority of cases;
however, fertility treatments, year x sludge rates, year
x crop history and sludge rates x crop history were
generally nonsignificant at $P < 0.05$. Therefore, this
report primarily addresses the main effects as they
impacted the availability of metals for uptake by tobacco
with treatments combined over years being the major
emphasis. In order to examine the effect of crop history
on the availability of metals with increased sludge
loading in the absence of supplemental S, the data from
the minus S treatments for the two years were combined in
an ANOVA which contained sludge rate and crop history as
the primary variables.

The results from Site II, which involved the growth
of soybeans on soils amended with the high metal sludge
from Baltimore, were examined as a randomized complete
block design with data combined over the two years of
1983 and 1984.

The data on metals for the tobacco and soybeans grown
at Site III, USDA-BARC, were examined independently of one
another since the two crops were grown simultaneously only
during one of the two years (1984). The results for each
crop were analyzed annually as a randomized complete block,
then combined ANOVA over years were performed. However,
due to the wide diversity of sludge types, the Site III
data were grouped into four sections: controls, limed
sludges, unlimed sludges, and Chicago sludge (Table 1).

The rainfall patterns at the three research sites
varied each year and influenced the growth of crops. As
would be expected, these differences caused some differences in relative quantities of metals taken up by a
given crop from one year to the next. However, the year
x sludge rates interactions generally were not significant
at $P < 0.05$ which suggest that the metal uptake patterns
were consistent across environments. For the purposes of
this report, the authors have attempted to focus on
general responses regarding metal uptake by crops grown
on sludge amended soils rather than on year to year
variability in the individual responses.

RESULTS AND DISCUSSION

Tobacco

Sludge application rates. Summaries of the leaf
metal contents in tobacco grown in 1983 and 1984 at site I
on soils amended in 1972 with digested sludge from the
Blue Plains treatment facility of Washington, D.C. are
contained in Tables 2 and 3. These results summarize
data from areas which were managed for soybeans or corn for
a decade following sludge treatments and which were not
subjected to elemental S treatments in 1983. There were
significant increases in the content of Zn, Cu, Ni and Cd
and trends for higher Mn and lower Fe levels with increased
sludge rates. Comparisons between the 0 vs. 224 Mg ha^{-1}
rate show Zn levels were increased 6-fold, Cu levels were
doubled and leaf Cd levels were tripled by the sludge
treatments. Increase in the content of Zn, Cu and Cd with
increased sludge rate exhibited quadratic rather than
linear responses which suggests that metal responses were
greater with the first increment of sludge addition than
from further additions. Levels of Pb in leaves were
generally unaffected by the sludge treatments although the
quantities of Pb added (15.5, 31 and 62 kg ha^{-1} Pb for the
56, 112 and 224 Mg ha^{-1}, sludge rates, respectively) were
substantial (Table 1).

Yearly means for metal contents averaged over sludge
rates and management sites are also listed in Table 2.
Reduced rainfall levels experienced in July, 1983, a period
of maximum plant growth for tobacco, caused significantly
lower amounts in leaves for all metals except Fe. Lead,

Table 2. Mean soil pH and metal contents in leaves of
tobacco grown in 1983 and 1984 on soils amended in 1972
with sludge from Washington, D.C. (minus S treatments
combined over management sites - UMD).

Sludge Rate (Mg ha^{-1})		Soil pH	Zn	Leaf Metal Contents (mg kg^{-1})					
				Cu	Mn	Fe	Pb	Ni	Cd
				Year Means					
	1983	5.6	135	25.9	114	399	2.62	2.77	9.04
	1984	5.4	162	29.1	173	368	3.74	3.81	9.98
	Stat. Sign	*	*	*	*	*	*	*	*
				Treatment Means					
0		5.8	57	15.9	120	419	3.06	3.50	4.01
56		5.6	147	31.0	146	410	3.16	3.20	8.85
112		5.3	220	30.5	164	377	3.49	3.23	11.67
224		5.2	300	32.8	195	379	3.32	3.90	13.52
	LSD (0.05)	0.3	48	4.9	NS	NS$^+$	NS	0.70	1.26

$^+$Significant at P $\leq$ 0.10.

*Significant at P $\leq$ 0.05.

Ni and Mn levels were reduced an average of 30 percent by
the drought conditions compared to 10 to 16 percent for
metals such as Cd, Cu and Zn. These reductions during
periods of moisture stress were likely caused by differences
in root extension patterns with the roots during 1983 being
more concentrated at deeper soil depths than during 1984.
The higher levels of moisture experienced in 1984 would
have favored greater root extension in the zones nearer
the soil surface which likely correspond to the zone of
highest metal enrichment. In 1972, the sludge treatments
were applied to the surface and incorporated in the upper
10-12 cm of soil using a disc. From discussions contained
in Frink and Hullar,[1] one would expect limited movement of
the metal ions in the silt loam soil. Also, the Beltsville
soil is known to possess a fragipan below the Ap horizon;
however, the drainage properties are such that the soil is
used in Southern Maryland for the growth of tobacco. The

Table 3. Mean soil pH and metal contents in leaves of tobacco grown in 1983 and 1984 on soils amended in 1972 with sludge from Washington, D.C. and managed for a decade of continuous soybeans or corn (minus S treatments – UMD).

| Sludge Rate (Mg ha^{-1}) | Soil pH | | Leaf Metal Contents (mg kg^{-1}) | | | | | | | | | | | | | |
| | Soy. | Corn | Zn | | Cu | | Mn | | Fe | | Pb | | Ni | | Cd | |
			Soy.	Corn	Soy.	Corn	Soy.	Corn	Soy.	Corn	Soy.	Corn	Soy.	Corn	Soy.	Corn
0	6.0	5.6	58	56	15.4	16.4	109	130	463	375	3.12	3.00	3.85	3.18	4.23	3.79
56	5.7	5.5	149	145	31.0	31.0	132	160	510	310	3.43	2.89	3.93	2.50	9.02	8.68
112	5.4	5.2	229	212	30.7	30.2	145	183	426	329	3.28	3.70	3.82	2.65	12.08	11.26
224	5.4	5.1	229	370	30.6	35.0	124	265	422	337	3.40	3.25	3.92	3.88	12.57	14.48
	---	---	---	---	---	---	---	---	---	---	---	---	---	---	---	---
Mean	5.6	5.4*	167	196*	26.9	28.1*	127	185*	455	337*	3.31	3.21	3.88	3.05*	9.47	9.55
LSD (0.05)‡	0.3		48		4.9		NS		NS		1.26		0.70		1.26	

*Site means were significantly different at p$\leq$ 0.05.

‡LSD value may be used for separation of sludge rate treatment means for both management sites.

reduced levels of Fe in 1984 would be consistent with lower
levels of soil oxygen which would accompany the higher
content of soil moisture. It should be stated that since
neither rooting patterns nor soil moisture levels were
monitored in this study, the data in Table 2 suggest that
these factors should be considered when using plants as
biological monitors for heavy metals in sludge amended
soils.

These data also confirm the necessity for examining
multiple environments when attempting to access the
bioavailability of metals by growing crops on sludge
amended soils. From the yearly means, this suggests that
variation in metal contents due to seasonal effects could
be as great or greater than some sludge treatment responses.
None of the leaf metal responses in the current study
exhibited a significant ($P < 0.05$) year x sludge rate
response.

Crop history. The effects of site management on
metal availability over a decade of continuous soybeans
or corn are illustrated in Table 3. The effects on the
soil are most vividly illustrated in the pH data where the
plots managed for soybeans average 0.2 to 0.4 units higher
than those managed for corn. These soil pH differences
caused significantly higher Zn, Cu, Mn and lower Fe and
Ni levels in the corn plots compared to the soybean plots,
especially at higher rates of application. Leaf Pb and
Cd levels were unaffected by the crop history treatments.

The results in Table 2 and 3 also illustrate the
effects of nonlimed sludge treatments on changes in soil
pH over time. Note that both crop history sites had
progressively lower pH's with increased sludge application
rates which resulted in the 224 Mg ha^{-1} treatments being
0.5 pH units lower than the nonsludged controls. The
lower pH values resulted from the decomposition of organic
matter and metabolism of NH_4^+ ions released from the sludge.
The full extent of pH changes in the absence of lime
additions during the previous decade are unknown. Records
show that supplemental liming materials were applied to
all plots on a regular basis initially then only the
sludge-amended soils were limed. Also, substantially
higher amounts of lime were applied with increased sludge
rate almost on an annual basis in attempts to keep pH
values uniform. Due to the higher C.E.C. in the sludge

amended soils, these soils will require larger amounts of
liming agents to induce desired pH changes and more frequent
lime applications to maintain pH values in the desired
range. The magnitude of the differences in lime requirements will likely be correlated with the quantity and
perhaps quality of sludge that was applied.

Soil acidity. The effect of applying supplemental S
in an effort to simulate the effects of future changes in
soil pH are illustrated in Table 4. The plots receiving S
were 0.3 pH units lower on average than the controls with
the 224 Mg ha^{-1} sludge rate averaging 0.4 units below
controls lacking S. These increased levels of soil
acidity stimulated greater plant uptake, i.e., bioavailability of Zn, Mn, Ni and Cd with the magnitude of the S
treatments being greatest for the 112 and 224 Mg ha^{-1} sludge
treatments. Some responses which merit particular comments
include those for Mn, Ni and Cd at the 224 Mg ha^{-1} sludge
rate. The 687 mg kg^{-1} Mn level may be considered as
exhibiting Mn toxicity in tobacco.[3] The 8.69 mg kg^{-1} Ni
and 19.25 mg kg^{-1} Cd levels are among the highest values
observed for tobacco and provides support for the general
conclusion to prohibit the growth of tobacco on sludge
amended soils due to enhanced risk of product contamination.[1,13]

The influence of soil pH on the content of metals in
leaves of tobacco in the absence of sludge treatments is
illustrated in data from the four control treatments at
USDA-BARC in 1984 and 1985 (Table 5). Although the
Christiana soil is not commonly used for growing tobacco
in Maryland, the results from the pH 4.9 and 5.3 nonsludged controls are within the supper ranges for values
found for tobacco grown in Maryland, especially for the
Collington and Monmouth soil series. If the results in
Table 5 for soil at 4.9 are typical for acid soils, it
becomes less difficult to ascertain the impact of the S
addition on enhancing the bioavailability of metals in the
sludge amended treatments shown in Table 4. It would
appear that the Mn and Ni responses in Table 4 were the
product of changes in soil pH rather than being caused by
increased sludge application. However, as will be
discussed in the next section, the responses cannot be
separated from sludge induced responses, especially when
the changes in pH were induced by sludge treatments as
shown in Tables 2 and 3. The data in Table 5 indicate that

Table 4. Mean soil pH and metal contents in leaves of tobacco in 1983 and 1984 on soil amended in 1972 with sludge from Washington, D.C. and treated with elemental S in 1983 (combined over management sites - UMD).

Sludge Rate (Mg kg⁻¹)	Soil pH		Leaf Metal Contents (mg kg⁻¹)												
			Zn		Cu		Mn		Fe		Pb		Ni		Cd	
	Cont.	+ S	Cont.	+ S	Cont.	+ S	Cont.	+ S	Cont.	+ S	Cont.	+ S	Cont.	+ S	Cont.	+ S
0	5.8	5.6	57	54	15.9	14.6	120	142	419	427	3.06	3.17	3.50	3.66	4.01	3.90
56	5.6	5.3	147	189	31.0	31.3	146	266	410	402	3.16	3.46	3.20	3.63	8.85	11.20
112	5.3	5.0	220	300	30.5	34.1	164	359	377	378	3.49	3.20	3.23	4.70	11.67	13.75
224	5.2	4.8	300	596	32.8	39.7	195	687	379	367	3.32	3.16	3.90	8.69	13.52	19.25
	---	---	---	---	---	---	---	---	---	---	---	---	---	---	---	---
Av.	5.5	5.2*	183	285*	27.5	29.9	156	363*	396	394	3.26	3.24	3.41	5.17*	9.03	12.03*
LSD (0.05)‡	0.3		134		3.5		228		NS		NS		2.19		2.12	

‡LSD values may be used for separation of sludge rate treatment means for controls and + S treatments.

* + S treatment means, averaged over sludge rates, were significant at P 0.05.

Table 5.   Summary of metal contents of leaves of tobacco
grown in 1984 and 1985 in control soils adjusted to higher
pH values with lime.   (USDA-BARC)

Treatments	Soil pH	Leaf Metal Contents (mg kg^{-1})					
		Zn	Cu	Mn	Cd	Ni	Pb
				Year Means			
1984	5.9	81	16.2	317	3.82	3.34	3.62
1985	6.0	56	12.1	175	3.72	2.03	2.37
Stat. Sign.	NS	*	**	*	NS	*	NS
				Treatment Means			
Control #1	4.9	126	17.1	670	4.61	5.26	3.21
Control #2	5.3	74	14.6	207	5.76	2.63	3.02
Control #3	6.2	46	14.0	78	2.73	1.89	2.84
Control #4	7.5	27	10.8	29	1.99	0.98	2.91
LSD (0.05)	0.2	16	2.8	156	0.75	1.46	NS

*,** Significant at P ≤ 0.05 and 0.01, respectively.

soil amendments, including limed sludge, which increased
the pH of the soil above 5.5 would substantially reduce
the metal contents in tobacco leaves.   Likewise, amendments
which ultimately lower the pH of the soil below 5.5 would
cause significant increases in the leaf contents for
several metals with sludge amended soils being among the
most responsive.

Sludge types and sources.   The tobacco data collected
were representative of a broad range of sludge types
(Table 1) with their chemical contents being influenced
by the areas served by the different waste water treatment
facilities.   Upon combining the data from Sites I and III,
tobacco was grown with increased rates of sludge application from two non-limed sludges (Blue Plains digested and
Annapolis heat treated), three limed sludges (Piscataway
limed digested, Piscataway limed raw, and Blue Plains
compost), and one sludge type of industrial origin
(Chicago's Nu-Earth).   Those with the lowest amounts of
Zn, Cu, Pb, and Cd include the two Piscataway sludges and
the Blue Plains compost (the three limed sources), with
Piscataway also having the lowest Ni contents (Table 1).
When compared with data for the maximum domestic sludge,
the Blue Plains digested sludge and the Annapolis heat
treated sludge exhibited intermediate levels for Zn, Cu,

Table 6. Summary of metal contents of leaves of tobacco
grown in 1984 and 1985 on soils amended in 1976 with heat
treated sludge from Annapolis, Maryland (USDA-BARC).

Treatments	Sludge Rate (Mg ha^{-1})	Soil pH	Leaf Metal Contents (mg kg-1)					
			Zn	Cu	Mn	Cd	Ni	Pb
Year Means								
1984	--	5.3	188	24.1	261	6.53	4.53	4.24
1985	—	5.4	116	21.9	138	5.07	2.76	2.22
Stat. Sign		NS	**	NS	**	**	**	NS
Treatment Means								
Control #1	0	4.9	126	17.1	670	4.61	5.26	3.21
Control #2	0	5.3	74	14.6	207	5.76	2.63	3.03
Annapolis	56	5.3	172	24.8	180	6.70	4.67	2.99
(Heat Treated)	112	5.1	205	24.9	156	6.81	4.13	2.98
	224	5.1	297	27.4	163	7.00	6.28	3.23
"	56	5.6	111	24.8	96	5.72	2.52	2.73
	112	5.6	116	24.3	79	5.15	1.64	2.43
	224	5.8	111	26.1	45	4.67	2.04	5.23
LSD (0.05)		0.2	50	3.0	111	1.10	1.82	NS+

*,**Significant at P ≤ 0.05 and 0.01, respectively.
+Significant at P≤ 0.10.

Pb, Ni and Cd. The Chicago Nu-Earth sludge exhibited Zn,
Cu, Ni and Cd levels in excess of those cited for domestic
sludges.[5]

    A summary of metal contents in tobacco from treatments
with Annapolis heat treated sludge maintained at soil pH
5.6 and 5.3 is contained in Table 6. These results vividly
illustrate the effect small increases in soil acidity have
on the availability of Zn, Cd and Ni to tobacco with
increased sludge application rates. At soil pH 5.6,
sludge rates up to 224 Mg ha^{-1} had no effect on Zn, Mn, Cd
or Ni but did show an apparent increase in Pb content. It
would appear that the threshold soil pH for determining
whether or not tobacco will be influenced by metals
applied with the Annapolis sludge is between pH 5.3 and
5.6. It would also appear that the availability of Cd in
the Annapolis sludge was much less than was noted for the
nonlimed Blue Plains sludge (Table 4).

Table 7.   Summary of metal contents of leaves of tobacco grown in 1984 and 1985 on soils amended in 1976 with limed digested and limed raw sludges from Piscataway and limed composted sludge from Blue Plains (USDA–BARC).

Treatment	Sludge Rate (Mg ha^{-1})	Soil pH	Leaf Metal Contents (mg kg^{-1})					
			Zn	Cu	Mn	Cd	Ni	Pb
Year Means								
1984	--	6.6	68	21.1	100	3.69	1.84	3.47
1985	--	6.5	43	14.0	53	3.37	1.40	2.32
Stat. Sign.		*	**	**	**	NS	NS	NS
Treatment Means								
Control #2	0	5.3	74	14.6	207	5.67	2.63	3.07
Control #3	0	6.2	46	14.0	78	2.73	1.89	2.84
Control #4	0	7.5	27	10.8	29	1.99	0.98	3.91
Piscataway	56	6.1	48	17.3	68	3.13	1.83	2.40
(Limed Digested)	112	7.1	39	18.5	37	2.84	1.13	2.87
	224	7.6	37	21.2	28	2.67	0.96	2.92
Piscataway	56	6.4	49	17.8	50	2.83	1.04	2.83
(Limed Raw)	112	6.9	42	19.0	38	3.12	0.94	0.90
	224	7.4	37	20.1	22	2.34	0.37	1.33
Blue Plains	56	5.6	81	19.9	126	4.95	2.17	3.38
(Limed Composted)	112	5.9	84	21.6	75	4.04	1.67	3.10
	224	6.5	65	20.4	40	3.68	1.08	2.82
LSD (0.05)		0.2	12	3.3	50	0.80	0.71	NS

*,** Significant at P ≤ 0.05 and 0.01, respectively.

The metal uptake responses by tobacco from the limed sludges applied at Site III are contained in Table 7.   In all three cases, the acidity levels were below pH 5.6 and were progressively reduced by additions of sludge which caused reductions in the concentrations for most metals except Cu.   However, when individual treatments are compared with appropriate pH controls, there were several instances where the levels of metals were increased by the sludge treatments.   For example, all of the Cu levels were above the controls, and the leaf Zn and Cd levels at 112 and 224 Mg ha^{-1} for Blue Plains compost were significantly higher than the control values.   It should be noted that all of the metal values listed in Table 7 were within normal background levels for Maryland tobacco, and the plant would

Table 8. Summary of metal contents of leaves of tobacco
grown in 1984 and 1985 on soils amended in 1978 with sludge
from Chicago, Illinois (USDA-BARC).

Treatments		Sludge Rate (Mg ha^{-1})	Soil pH	Leaf Metal Contents (mg kg^{-1})					
				Zn	Cu	Mn	Cd	Ni	Pb
Year Means									
	1984	--	5.8	173	25.2	142	34.9	6.17	3.73
	1985	--	6.0	96	22.0	63	23.1	3.05	2.24
	Stat. Sign.		NS	**	*	*	*	**	NS
Treatment Means									
Control #2		0	5.3	74	14.6	207	5.76	2.63	3.02
Control #3		0	6.2	46	14.0	78	2.73	1.88	2.84
Chicago		50	5.5	225	30.7	123	49.10	7.98	3.01
(Nu-Earth)		100	5.8	242	30.0	88	56.00	9.13	2.90
"		50	6.2	106	24.7	82	25.00	2.77	3.08
		100	6.5	117	27.5	37	35.4	3.29	3.05
	LSD (0.05)		0.2	52	2.9	69	10.6	2.4	NS

*,** Significant at P $\leq$ 0.05 and 0.01, respectively.

therefore not be considered as being contaminated by heavy
metals from sludge.

    The bioavailability of metals for uptake by tobacco
when grown on a sludge of industrial origin (i.e., Chicago's
Nu-Earth) is illustrated in Table 8. Significant differences were observed with increased sludge application rate
for all metals examined except Pb; however, the Mn results
appear to be mediated by soil pH rather than by sludge rate.
Regarding the interaction of sludge rate with pH, these
results differ from the previously discussed results from
Site III in that the significant increases in leaf Zn, Cu
and Cd contents were found with increased sludge rate at
pH 6.2-6.5. In Table 6, the Zn, Cd and Ni were largely
controlled at pH 5.6. For the Chicago sludge, the Ni
applied in sludge was generally unavailable at pH 6.2-6.5
whereas the Cd, Zn and Cu showed increased levels of
availability with increased sludge rate. However, when
the soil acidity increased to pH 5.5-5.8, the concentrations of leaf Zn, Cd and Ni were typically twice the
values found at the lower acidity levels.

The differences in bioavailability of metals observed among the several sludge types, especially among the unlimed sludges, can be explained by examining the differences in metal contents in the sludge (Table 1). The sludge from the industrial centers had much higher concentrations of Zn, Cu, Pb, Ni and Cd than the Annapolis and Blue Plains sludges. In addition, the Fe levels in the Annapolis sludges, which may help to inhibit plant metal uptake, were 2 to 3 times greater than that from the other sources. It would appear that pH alone is not sufficient to control the bioavailability of certain metals from industrial centers and that lime or supplemental Fe in sludge can have a substantial impact on mediating the bioavailability of metals from non-industrial sources. The liming agents prevents the soil from becoming highly acidic during the breakdown of organic matter. Questions which have not been answered by the present research concern the bioavailability of metals for metal sensitive crops grown on limed sludge soils once the soil pH returns to background levels - i.e., pH values between 4.7 and 5.3 for acid soils in Maryland. Under poor management, these soils would likely respond to pH as illustrated by the results in Tables 2-4.

Soybeans

The data for soybeans grown on soils amended with heat-treated sludge from Annapolis, Maryland seven years after application are summarized in Table 9. Compared to the tobacco results for similar treatments (Table 6), the amounts of metal in soybeans were 0.5 to 0.05 times lower in the bean shoots, except for Ni values, which tended to be higher than values found in tobacco. Values of Pb in shoots were not examined in the soybean study. As was also found with tobacco, year to year variation was quite large and soil pH played an equally important role in mediating the uptake of metals by soybeans. The contents of Zn in shoots, for example, exhibited progressively higher levels with increased sludge application with the response at pH 5.6 being about double the value at pH 6.2. However, the amounts of Cd were generally unchanged with applied sludge at pH 5.6 but did show significant change at the higher sludge rate at pH 6.2. The amount of Cu in shoots was largely unaffected or reduced by increased sludge rate rather than the consistent increases observed for tobacco. Levels of Ni in shoots were increased by the

Table 9. Summary of metal contents of shoots of soybean grown in 1983 and 1984 on soils amended in 1976 with heat treated sludges from the Annapolis, Maryland treatment plant (USDA-BARC).

Treatments	Sludge Rate (Mg ha^{-1})	Soil pH	Shoot Metal Contents (mg kg^{-1})				
			Zn	Cu	Mn	Cd	Ni
			Year Means				
1983	--	5.9	80	7.5	62	0.26	7.6
1984	--	5.9	64	6.5	43	0.22	3.7
Stat. Sign		NS	**	*	**	NS	**
			Treatment Means				
Control #1	0	5.6	55	7.2	105	0.33	5.9
Control #2	0	6.2	36	7.9	50	0.11	4.6
Annapolis	56	5.6	84	7.9	46	0.23	8.8
(Heat-treated)	112	5.7	100	6.6	52	0.34	6.1
	224	5.7	130	6.0	48	0.31	7.2
"	56	6.2	51	7.3	40	0.19	5.4
	112	6.0	60	6.5	46	0.17	3.8
	224	6.2	58	6.7	31	0.24	3.4
LSD (0.05)		0.3	20	0.7	19	0.09	2.0

*,** Significant at P ≤ 0.05 and 0.01, respectively.

application of 56 Mg ha^{-1} sludge at pH 5.6 but not at pH 6.2; however, increased sludge application above 56 Mg ha^{-1} tended to reduce the shoot Ni content. Therefore, the bioavailability of Cu and Ni for soybean grown on soil amended with non-limed sludge was clearly different than was found for tobacco.

Table 10 shows that increasing the pH from 5.6 to 7.3 produced significant reductions in the uptake of Zn, Mn, Cd and Ni and increased uptake of Cu by soybeans. The increased application of limed sludge from Piscataway, whether digested or raw, produced few significant changes in any of the trace metals examined in soybeans; however, Zn and Cd levels were increased by applications of Blue Plains composted sludge.

The effect of industrial sludges on the uptake of metals by soybeans are contained in Tables 11 and 12 for the high-metal Chicago and Baltimore sludges (Table 1).

Table 10. Summary of metal contents of shoots of soybeans grown in 1983 and 1984 on soils amended in 1976 with limed digested and limed raw sludges from Piscataway and limed composted sludge from the Blue Plains treatment plants (USDA-BARC).

Treatments	Sludge Rate (Mg ha^{-1})	Soil pH	Shoot Metal Contents (mg kg-1)				
			Zn	Cu	Mn	Cd	Ni
Year Means							
1983	--	6.7	43	7.7	52	0.19	5.4
1984	--	6.7	30	8.0	37	0.11	2.2
Stat. Sign.		NS	**	NS	**	*	**
Treatment Means							
Control #1	0	5.6	55	7.2	105	0.33	5.9
Control #2	0	6.2	36	7.9	50	0.11	4.6
Control #3	0	7.3	25	8.6	28	0.08	4.2
Piscataway	56	6.4	38	8.0	45	0.08	4.7
(Limed Digested	112	7.2	34	8.3	38	0.19	4.2
	224	7.7	29	8.5	33	0.07	2.6
Piscataway	56	6.5	29	7.9	44	0.12	2.9
(Limed Raw)	112	7.1	28	7.8	34	0.10	2.2
	224	7.6	30	8.3	23	0.08	2.0
Blue Plains	56	6.0	46	7.5	53	0.20	5.3
(Composted)	112	6.3	45	7.0	43	0.23	3.4
	224	6.6	42	7.6	34	0.20	3.4
LSD (0.05)		0.3	9	0.6	17	0.12	1.6

*,** Significant at P $\leq$ 0.05 and 0.01, respectively.

Significant increases in shoot Zn, Cu, Cd and Ni were found with increased rates of application of Chicago and Baltimore sludge. Compared to the total quantities of metals applied in each of these high-metal sludge treatments, the levels of response were one order of magnitude lower than was observed for tobacco and suggest that soybeans have a very low sensitivity to metals applied in sludge. Therefore, soybeans could be safely grown on soils amended with industrial sludges at very high rates without problems with metal toxicity.[5]

Table 11. Summary of metal contents of shoots of soybean grown in 1983 and 1984 on soils amended in 1978 with municipal sludge from Chicago, Illinois (USDA-BARC).

Treatments		Sludge Rate (Mg ha^{-1})	Soil pH	Shoot Metal Contents (mg kg^{-1})				
				Zn	Cu	Mn	Cd	Ni
				Year Means				
	1983	--	6.2	73	7.6	67	1.64	8.2
	1984	--	6.2	58	7.3	42	1.32	4.6
	Stat. Sign.		NS	**	*	*	*	**
				Treatment Means				
Control #1		0	5.6	55	7.2	105	0.33	5.9
Control #2		0	6.2	36	7.9	50	0.11	4.6
Chicago, IL.		50	5.5	79	7.1	49	2.15	7.7
(Nu-Earth)		100	6.4	90	7.0	38	2.79	9.0
"		50	6.5	64	7.8	50	1.68	5.6
		100	7.0	67	7.8	34	1.84	5.8
	LSD (0.05)		0.3	10	0.6	22	0.46	1.8

*,** Significant at P ≤ 0.05 and 0.01, respectively.

Table 12. Effect of sludge rates on the microelement composition of shoots of soybean grown on soil amended with sludge from Baltimore, Maryland in 1976.

Sludge Rate (Mg ha^{-1})	Soil pH	Microelements (mg kg^{-1})				
		Zn	Cu	Mn	Cd	Ni
		Yearly Means				
1983	6.6	90	8.6	44	0.16	3.9
1984	6.6	97	9.0	39	0.14	4.4
Stat. Sign.	--	*	*	*	NS	NS
		Treatment Means				
0	6.9	25	8.4	50	0.07	2.7
56	6.6	92	8.7	39	0.16	4.2
112	6.4	150	9.3	36	0.23	5.6
LSD (0.05)	---	20	0.2	4	0.04	0.8

*Significant P ≤ 0.05.

CONCLUSIONS

The application of municipal sludges from several
sources, and representing a variety of technologies, was
observed to influence significantly the bioavailability
of metals a decade after the sludge treatments. Metals
which exhibited the largest changes in availability
included Zn, Cu, Cd and Ni with Fe, Mn and Pb being rarely
affected. Non-limed sludges were observed to rapidly
decrease the soil pH thus making the metals more available.
Limed sludges were observed to increase the soil pH to
values in excess of 7.0 which greatly reduced the availability of metals for crop uptake. Industrial sludges
containing high concentrations of heavy metals resulted
in the highest levels of metals in plants.

The manner in which the soils were managed following
sludge application had a large influence on the uptake of
metals by plants with soil pH having a central role in
regulating the bioavailability of metals. Soils allowed
to become acid (i.e., pH below 5.5) had significantly
higher levels of available metals than soils managed
above pH 6.0.

The type of crops grown on sludge amended soils
exhibited significantly different levels of metals in
their vegetative parts with tobacco being more sensitive
than soybeans. There was substantial risk from Cd and Ni
contamination of tobacco leaves when grown on sludge amended
soils having pH values below 6.0. The levels of metals in
soybean shoots in most cases were below levels which merit
concern with regards to product contamination.

Environmental conditions under which the crops were
grown had a significant impact on the levels of metals
accumulated in the leaves and shoots in both crops examined.
Extended drought was observed to lower the vegetative
metal contents in excess of 30 percent for certain metals.

ACKNOWLEDGMENT

The authors wish to acknowledge the assistance of
several individuals: Dr. C.G. McKee, who assisted with the
tobacco studies; Dr. Rufus Chaney of USDA-BARC who established the Hayden farm (Site III) treatments; and Dr.

Morris Decker, who managed the University of Maryland's Sites I and II following sludge application in the early 1970's.

Financial assistance for these studies were provided by the R.J. Reynolds Tobacco Co., the Tobacco Laboratory at USDA-BARC and the Maryland Agricultural Experiment Station. This is Scientific Article A-4469 Contribution No. 7461 of the Maryland Agricultural Station, Department of Agronomy, College Park, Maryland 20742.

REFERENCES

1. FRINK, C.R., T.L. HULLAR. 1984. Criteria on recommendations for land application of sludges in the northeast Pennsylvania State University. Penn. Agric. Exp. Sta. Bull. 851, p. 94.
2. U.S. ENVIRONMENTAL PROTECTION AGENCY. 1979. Criteria for classification of solid waste disposal facilities and practices. Federal Register 44 (179): 55438-53464.
3. BELL, P.F. 1986. M.S. Thesis. Residual effects from land applied sewage sludge on the agronomic, chemical and physical properties of Maryland tobacco. University of Maryland, College Park, Maryland.
4. COUNCIL FOR AGRICULTURE, SCIENCE AND TECHNOLOGY. Effects of sewage sludge on the cadmium and zinc content of crops. Council for Agric. Sci. Tech. No. 83, Ames, Iowa, p. 7.
5. CHANEY, R.L., P.M. GIORDANO. 1977. Microelements as related to plant deficiencies and toxicities. In Soils for Management of Organic Waste and Waste Waters. (L.F. Elliott and F.J. Stevenson, eds.), Amer. Soc. Agron., Madison, Wisconsin, pp. 234-279.
6. HECKMAN, J. 1985. M.S. Thesis. Effect of sewage sludge on soybean heavy metal accumulation and symbiotic nitrogen fixation. University of Maryland, College Park, Maryland.
7. SOMMERS, L.E. 1977. Chemical compositions of sewage sludges and analysis of their potential use as fertilizers. J. Environ. Qual. 6: 225-232.
8. SOON, Y.K., T.E. BATES, J.R. MOYER. 1980. Land application of chemically treated sewage sludge: III. Effects on soil and plant heavy metal content. J. Environ. Qual. 9: 497-504.

9. CHANEY, R.L., S.B. STERRETT, M.C. MORELLA, C.A. LLOYD.
    1982. Effect of sludge quality and rate, soil pH
    and lime on heavy metal residues in leafy vege-
    tables. In Proceedings of the Fifth Annual Madison
    Conference on Applied Research Practice on Municipal
    and Industrial Waste, University of Wisconsin,
    Madison, Wisconsin, pp. 444-458.
10. COREY, R.B., R. FUJII, L.L. HENDRICKSON. 1981.
    Bioavailability of heavy metals in soil-sludge
    systems. In Proceedings of the Fourth Annual
    Madison Conference on Applied Research Practice on
    Municipal and Industrial Waste, University of
    Wisconsin, Madison, Wisconsin, pp. 449-465.
11. BAKER, D.E., A.M. WOLF. 1984. Cadmium. In Criteria
    on Recommendations for Land Application of
    Sludges in the Northeast. (C.R. Frink, T.I.
    Hullar, eds.), Pennsylvania State University,
    Penn. Agric. Exp. Sta. Bull. 851, pp. 43-49.
12. DEAN, R.B., M.J. SUESS. 1985. The risk to health of
    chemicals in sewage sludge applied to land.
    Waste Man. Res. 3: 251-278.
13. ELINDER, C.-G., T. KJELLSTROM, C. FRIBOG, B. LIND,
    L. LINNMAN. 1976. Cadmium in kidney cortex,
    liver and pancreas from Swedish amtopsies.
    Arch. Environ. Health 23: 292-302.
14. MACLEAN, A.J. 1976. Cadmium in different plant
    species and its availability in soils as influenced
    by organic matter and additions of lime, phosphorus,
    cadmium, and zinc. Can. J. Soil Sci. 56: 129-138.